Birkhäuser

Mathematics and the Built Environment
Volume 2

Series editor

Kim Williams, Kim Williams Books, Torino, Italy

More information about this series at http://www.springer.com/series/15181

Brian Wichmann · David Wade

Islamic Design: A Mathematical Approach

Birkhäuser

Brian Wichmann
Woking, UK
brian.a.wichmann@gmail.com

David Wade
Lanidloes, UK
davidgwade@gmail.com

ISSN 2512-157X ISSN 2512-1561 (electronic)
Mathematics and the Built Environment
ISBN 978-3-030-09923-7 ISBN 978-3-319-69977-6 (eBook)
https://doi.org/10.1007/978-3-319-69977-6

Printed on acid-free paper

This book is published under the imprint Birkhäuser, www.birkhauser-science.com by the registered company Springer International Publishing AG part of Springer Nature
The registered company address is: Gewerbestrasse 11, 6330 Cham, Switzerland

Foreword

There will always be an element of mystery about the ways in which cultures manage to stamp their character on their respective decorative styles. Celtic art, for instance, has its own unmistakable characteristics, as do the ornamental arts of China and the various pre-Columbian cultures. Islamic decorative art, particularly in its mature forms, is immediately recognizable in the same way, but is all the more remarkable for its duration and the sheer extent of its geographical spread. The Islamic decorative canon, which is comprised of calligraphy, foliated arabesque and geometrical pattern, can be found throughout the Islamic world—which is to say, from the Atlantic to the borders of China—and has endured for hundreds of years; it clearly reflects some essential aspect of the Islamic world view. Each element of this decorative tradition underwent its own processes of development and refinement, but this book focuses on the latter component, namely the complex geometric patterning that has become so firmly associated with Islamic culture.

Prior to the modern era, the Islamic and European spheres tended to go their own separate ways and it was not really until the nineteenth century that the architectural and artistic achievements of the Islamic world began to be recognized in the West. The Alhambra Palace in southern Spain was among the first of Islamic monuments to receive serious attention from European scholars and other visitors. As a result, many studies were published during this period, notably those by the British architect Owen Jones, see (Jones 1856); these had the effect of stimulating a wider interest in Islamic ornament. Before this, the large-scale archaeological expedition ordered by Napoleon after his conquest of Egypt had given rise to an enduring interest in France for all things Egyptian—including its Islamic art and architecture. In time, this grand project led to various important studies of the subject, notably those of two Frenchmen Prisse d'Avennes and Jules Bourgoin, who went on to publish remarkable volumes on Islamic art.

There were, of course, other points of contact between the European nations and the Islamic world at this time—the British rule in India, French expansion into Algeria and Morocco and Russia incursions into Central Asia, for instance. All of these, essentially imperialist, ventures had the redeeming side effect of generating interest in the cultural achievements of the invaded territories, including investigations into the unique Islamic genre of geometric art that is the subject of this book. These achievements, some of which were created at a time of Islamic decline, went on to make an indelible impression on the arts and crafts of the West.

The evolution of Islamic decorative art itself, as with all other form of artistic expression, owes something to its immediate predecessors (in this case the decorative forms of the Late-Classical Syro-Roman, Byzantine and Sassanian Persian eras), but it seems also to have been influenced by a rather older, mathematical tradition, namely that of Classical and Hellenistic Greek geometry. Indeed, the spirit of Euclid hovers around these designs and undoubtedly contributes to their unique quality of geometrical/mathematical complexity.

Part I of this book provides an overview of the Islamic world from the time of Mohammed to the modern era. This includes the artistic, scientific and philosophical aspects of the culture that gave rise to its unique geometrical decorative forms. Part II gives a formal analysis of a great range of these patterns using modern mathematical techniques supported by computer graphics. In this way, the constructs and vision of the original artist/craftsmen can be reconstructed using today's technology.

Notes on Sources and Presentation

This book derives much of its material from the database found on the website http://www.tilingsearch.org which is obviously available for those interested in further data. This website is continuously updated as more information about Islamic geometric patterns becomes available. If the reader experiences any difficulty in contacting this site, it can be accessed via the British Library archive.

Most of the photographic material relating to geometric patterns in this book comes from the photo-archive available in the website http://patterninislamicart.com which contains over 3600 images (excluding published material), drawn from many parts of the Islamic world.

The colours used in the explanatory diagrams in this book are from the original artefact when appropriate or changed to emphasize the construction of the pattern. When the dates of a pattern can be determined, these are presented here as AD (or CE, Common Era); the AH date (in the year of the *Hegira*) has been left out for ease of presentation.

The illustrations in Part I are mainly photographs, while those in Part II are mainly computer graphics.

Acknowledgements

A key acknowledgement goes to Tony Lee who produced much of the background information upon which the website (www.tilingsearch.org) is based. We would also like to thank Tony for all his help and advice in adapting his background information for use in Part II of the book.

Our thanks to Bernard O'Kane, Professor of Islamic Art and Architecture, The American University in Cairo, who read a draft of this book, provided excellent photographs and answered many queries about Egypt.

We thank all the people who provided valuable assistance and information during the construction of the website (www.tilingsearch.org). We thank them for their contributions which have subsequently been used in the preparation of this book.

We are grateful to the staff of the Victoria and Albert Museum, whose help has been invaluable to this study of Islamic design, especially the National Art Library.

Thanks to two academics who provided photographs from the National Library and Archives of Egypt: Dr. Sabah E. Sulaiman, Professor (A) of Architecture, Faculty of Engineering, Ain Shams University, Cairo, Egypt, and Dr. Ahmed Ali Elkhateeb, Department of Architecture, Faculty of Engineering, Ain Shams University, Cairo, Egypt.

Additional photographs have been invaluable for this book and were contributed by Nick Crossling, Richard Henry, Goossen Karssenberg, Brian McMorrow and Ayman Soliman.

Three websites have also been particularly useful: (http://archnet.org, www.islamic-art.org, www.discoverislamicart.org).

Thanks to Anne Comment (Editorial Assistant-Mathematics at Springer) for all her support in the publication of this book.

Contents

Part I Cultural Context

1 A History 3
- 1.1 The Revelations 3
- 1.2 Early Conquests 4
- 1.3 Islam and the Civilisations of the Ancient World 4
- 1.4 The Umayyads and Damascus 5
- 1.5 The Abbasids and Baghdad 5
- 1.6 The Disintegration of the Caliphate 6
- 1.7 The Sunni Revival 7
- 1.8 The Mongol Cataclysm 8
- 1.9 The Islamic Revanche: The Ilkhanids and Timurids 10
- 1.10 The Early-Modern Islamic States 11
- 1.11 Spain and the Maghreb 13
- 1.12 Imperialist Encroachment and Islamic Decline 14

2 The Scientific Contribution 17
- 2.1 The Greek Connection 17
- 2.2 The Pythagorean/Platonic Tradition 19
- 2.3 Philosophy in Islam 20
- 2.4 Religious Revival, Scientific Retreat 21
- 2.5 Continuity and Transmutation 22

3 The Religious Dimension 25
- 3.1 'There Is No God but God' 25
- 3.2 The Qur'ān 26
- 3.3 Traditions (Hadith) 27
- 3.4 Islamic Identity and the Image Controversy 27
- 3.5 Mizan, Symmetry and Cosmic Equilibrium 28
- 3.6 The Islamic Paradise (al-Djanna) 29

4 The Evolution of Style 31
4.1 Consistency and Variety 31
4.2 Symmetry and Geometry 32
4.3 Earlier Islamic Art 34
4.4 Baghdad, Samarra and Political Fragmentation; 8th–10th Centuries 34
4.5 Stylistic Maturity; 11th–12th Centuries 35
4.6 The Islamic Decorative Canon 36
4.7 The Craft Group 37
4.8 The Role of Paper in Islamic Art and Architecture 38

5 Materials and Media 41
5.1 Stonework 41
5.2 Brickwork 42
5.3 Ceramic Tiles and Ceramic Mosaic 43
5.4 Stucco/Plasterwork 45
5.5 Woodwork 45
5.6 Book Illumination 47
5.7 Court Patronage and Cottage Industry 48

6 Countries and Regions 51
6.1 Egypt 51
6.2 India 51
6.3 Iran 53
6.4 Morocco 53
6.5 Spain 54
6.6 Syria 55
6.7 Central Asia 56
6.8 Turkey 58

Part II Mathematical Analysis

7 Introduction 63
7.1 Geometry 63
7.1.1 Rosettes 64

8 A Worked Example 69

9 An Octagonal Set 75

10 Octagonal Tiles with a Large Star 87
10.1 16-Pointed Stars 87
10.1.1 Adding Octagonal Tiles 88
10.2 24-Pointed Stars 93
10.3 A 32-Pointed Star 99

11 Lines and Edges 103
11.1 Strict Interlacing 105
11.2 Borders 109
11.3 Banding 109
11.4 Irregular Borders 112
11.5 Interlace Path Tracing 114

12 Decagonal Patterns 117
12.1 A Simple Tile Set 117
12.2 Variable Diamond Patterns 119
12.3 Fixed-Size Diamond Patterns 121
12.4 Adding a 5-Pointed Star 121
12.5 Another Simple Tile Set 124
12.6 Some Erratic Patterns 128

13 What Is Correct? 129
13.1 An Example from the Alhambra 129
13.2 Bourgoin Plate 32 131
13.3 Bourgoin Plate 75 132
13.4 An Egyptian Qur'ān 135
13.5 Conclusions 136

14 6-Fold Delights 137
14.1 Roman Influences 137
14.2 Simple Islamic Designs 139
14.3 Simple Islamic Designs Without Stars 139
14.4 Simple Islamic Designs with a Single Rosette 140
14.5 Designs with Two Rosettes 142
14.6 Some More Complex Designs 146
14.7 A Motif with Two Symmetries 147

15 Two-Level Patterns 151
15.1 An Iranian Example 151
15.2 An Indian Example 152
15.3 A Decagonal Example 153
15.4 The Topkapı Scroll 155
15.5 A Complex Issue to Resolve 157
15.6 A Complex Pattern: Darb-I Imam Shrine 160
15.7 A Complex Pattern: Seyyed Mosque 160

16 Two Mamluk Masterpieces 167
16.1 The Trigonometric Method 169
16.2 The Manual Method 172
16.3 A Comparison 175
16.4 A Second Mamluk Masterpiece 177

17 Conclusions 183
17.1 Approximate Methods 183
17.2 Ruler and Compass 184
17.3 Use of Mathematics 185
17.4 The Geometric Time-Line 186
17.5 Pattern Type by Region 188
17.6 A Summary of Methods 189
17.6.1 Complete Analysis 190
17.6.2 Incomplete Analysis 190

Appendix A: The Symmetry of a Tiling 191

Appendix B: Key Dates 211

Appendix C: Glossary 213

Appendix D: Copyright 215

References 217

Index 221

Part I
Cultural Context

By David Wade

The Great Mosque of Damascus, one of the oldest and most important mosques in the Islamic world, has a long and checkered history. Important to both the Sunni and Shia traditions, it presents decorative features from many periods, going back to the earliest, Umayyad dynasty when the influence of late classical tradition prevailed. This mihrab, of which the above is a detail, represents a virtual tour de force of geometric patterning. Dating from the medieval period, it has been carefully and lovingly restored, as befits such a masterpiece of architectural decoration.

Chapter 1
A History

The creation within the space of a single century of a vast Arab Empire stretching from Spain to India is one of the most extraordinary marvels of history.

J.J. Saunders, A History of Medieval Islam (Saunders 1965, p. 39)

1.1 The Revelations

The bare historical facts concerning the origin of the religion of Islam are these—The Qur'ān, the Holy Book of Islam, began to be revealed to the Prophet Muhammad when he was forty years old, circa 610. These revelations continued until his death in 632, by which time their strict monotheistic message had become so widely accepted that the new religion had largely supplanted the ancient paganism of the Arabian Peninsula. The rapid, widespread acceptance of Islam also introduced a quite new sense of unity among the various Arabian tribes. For a list of key dates, see Appendix B.

This unity and sense of religious purpose was disturbed by the death of Muhammad, but after a short period of uncertainty his successors (in Arabic, *Khalifah*) continued his mission, with an expansive programme of conversion, diplomacy and military tactics. The first Caliph, Abu Bakr, defeated various tribal rebellions that followed the Prophet's death and managed to hold the Islamic community together. In 634 he called for a *jihad* (Holy War) to spread the Faith beyond the Arabian borders. In this enterprise the new religion was to prove extraordinarily successful.

B. Wichmann and D. Wade, *Islamic Design: A Mathematical Approach*, Mathematics and the Built Environment 2,
https://doi.org/10.1007/978-3-319-69977-6_1

1.2 Early Conquests

The decision to expand the 'mandate of Allah' outwards, to the north and east, inevitably meant that the Muslims would confront the great Empires of Byzantium and Persia. In fact the timing, from their point of view, was propitious since these rival civilisations had been engaged in an exhausting war with each other for some twenty-five years (603–628), as a result of which the resources of both were seriously depleted. The military successes of the upstart Arabs against the military might of both the Byzantine and Persian armies were quite remarkable. Damascus, the most important town in the Byzantine Middle-East, had capitulated by 635, the rest of Syria by 636. For the native Syrians, both Christians and Jews, this change of rulership was not altogether unwelcome, since it represented a release from Byzantine tyranny—another reason for the success of Islam in this region. Mesopotamia fell to the Arab forces in 637 and Egypt in 639. In the following decades Persia, Turkestan, the Indus Valley, North Africa and Spain would follow. Remarkably, the Islamic world was largely delineated within a century of the death of the Prophet. Naturally, the rapidity and scale of these early conquests were felt by the early Muslims as a vindication of the rightfulness of their beliefs; God was clearly on their side.

1.3 Islam and the Civilisations of the Ancient World

There had been great nomadic movements before and there were others after the Arab conquests, but the important difference between the Islamic expansion and all others was precisely the intense collective conviction that was inspired by their newly-found faith. Fired as they were by religious zeal, these essentially unlettered nomads were able to withstand a complete absorption into the more sophisticated civilisations that had fallen into their lap. From the beginning, their mission was not merely destructive; they had a mission to reinvigorate the 'Abrahamic' religion where it already existed (in the Hellenised, Christian remnants of the Eastern Roman Empire), and introduce it where it did not (in the Persian Sassanid Empire, and beyond). So although the wealth and technical achievements of the civilisations that they conquered made a deep and lasting impression they were able to resist the ideological and religious values that they encountered—in fact they were supremely confident of the superiority of their own religious beliefs. However, they had less resistance to more general cultural influences. To retain their hold on the conquered territories, the Muslims had to adopt many of their civilised ways. They proved very adept at this, largely by retaining the existing administrative institutions (including taxation and bureaucracy). In short, they quickly learned everything that sedentary civilisations had to teach them that was necessary to consolidate their conquests.

1.4 The Umayyads and Damascus

Muhammad had made Mecca the religious centre of the Islamic world (a role that it still fulfils), but its geographical position made it too remote a location from which to rule the recently acquired territories. It was inevitable that the Islamic administrative centre of gravity would shift northwards. In 661, the role of Caliph was taken over by Mu'awiya, an able general who had led the campaigns against the Byzantines and who had governed Syria for some twenty years. Mu'awiya belonged to the Meccan clan of Omayya, so he and his descendants were known as the Umayyad, and they would continue to rule the new Islamic Empire from the ancient town of Damascus for almost a century. This period of Islamic history saw the extraordinary expansion referred to in the heading quote. There was a downside to this campaign; it has been said that under the Umayyad Caliphate Islam grew as a power, but decayed as a religion.

The military conquests certainly continued at a staggering pace; by 710 Muslim troops had conquered the whole of North Africa and were crossing the Straits of Gibraltar to Spain; by 714 it too had succumbed. At the very same time, the Arabs, having completely overcome Persia, were pushing east to Central Asia, which capitulated in 712, as did the Indus Valley in 713 (both of which have retained an allegiance to Islam ever since).

At the centenary of the Hegira then, a centrally-controlled Islamic Empire ruled over territories that stretched quite literally from Spain and the Atlantic coast of Africa to India and the borders of China. As it turned out, these were to be the limits of Islamic territorial expansion for many centuries to come. In the process of Islam's transformation from a religious movement into an imperial one it had lost much of its early spirit of group solidarity and to some extent its sense of missionary purpose. Serious rifts and dissatisfactions began to appear in the body politic, and these tended to be directed towards the ruling Dynasty. The Umayyad Emperors had been very successful in taking over the administrative practices of the Greeks and Persians, but they had also adopted their somewhat despotic manners, becoming luxurious, remote and autocratic as a consequence. Despite their unprecedented achievement in creating such a vast Empire, the Umayyads provoked many discontents. The dissenting factions finally came together in 750, when the Umayyads were overthrown by a revolution.

1.5 The Abbasids and Baghdad

In reality, the Umayyads had never been generally popular; the contrast between their Imperial manners and the austere lives of the first Caliphs was just too obvious. Towards the end the dynasty was also bedevilled by its own internal feuds and jealousies. Once the different strands of opposition to their rule was organised into armed resistance, their regime collapsed. Much of the resentment of Umayyad rule had came from the *mawali*, the non-Arab converts to Islam, but it was another Arab dynasty that benefited from the revolution.

The Abbasid, who took over the Caliphate, promised a government that would be based on true Muslim principles. They were descended from the Prophet's family, which gave their claim to the Caliphate greater legitimacy in the eyes of the pious. But their promises of a *Dawla*, a 'New Order' for Islam, harking back to the early days, were abandoned soon after they were established in power. In reality, the Caliphate under the Abbasids was steadily Persianised. This began with the removal of the seat of government to a new capital, Baghdad, which was built near the ruins of Ctesiphon in 762. Here the Caliphs began to adopt the sort of sacred absolutism that characterised the Kings of the Ancient World, of Ninevah, Babylon and Persia. The Abbasids were soon portraying themselves as the 'Shadow of God on Earth'. Their dynasty, however, become one of the most enduring in Islamic history, the Abbasid Caliphate itself lasting until 1258.

The early Abbasid centuries were the period in which many of Islam's abiding Islamic cultural values were formed. In Baghdad the remnants of the Classical traditions of Rome and Greece were combined with those of Persia and India to form a brilliant new synthesis. In this way Islam's originality as a religion was translated into a confident, highly civilised culture with its own distinctive ethos and social values. It developed a system of Law (derived from the Qur'ān and the *Hadith*), and an enthusiasm for science and mathematics. The formation of a distinctive Islamic style in art, architecture and music also dates from this time. Sadly, much of the details of these interesting developments were lost to history, largely as a result of the terrible destruction that Baghdad was later to endure.

The Golden Age of Baghdad is often associated with the reign of the fifth Abbasid Caliph, *Harun al-Rashid* 786–806. Among other achievements of his time (which were important, not only for Islam but for the entire civilised world) was the concerted attempt to translate the remaining corpus of Greek scientific and other material into Arabic (see Chap. 2).

1.6 The Disintegration of the Caliphate

Brilliant though it was, the Abbasids suffered the same difficulties in maintaining political unity over their vast territories that had bedevilled the Umayyads. The problems, once again, were both internal and external. After Harun al-Rashid's death, quarrels between his sons over the succession led to civil war, the factions dividing along ethnic lines. The victor of this particular feud was Harun's son Ma'mun, who became a renowned patron of the arts and science. In 828 Ma'mun, continuing his fathers lead, founded the *Dar-al-Hikhma*, the famous 'House of Wisdom', which was dedicated to translating Greek scientific works. Unfortunately, his enthusiasm for the knowledge of the older civilisations was not matched by the sort of political instincts necessary to run a vast Empire. With so many different ethnic loyalties and traditions of belief within the Empire there were always bound to be rivalries and separatism. In fact from the early 9th century, the Abbasid Caliphate gradually began to lose its grip, although the decline was uneven, and there were periods of recovery and reconsolidation. As well as territorial disputes the Abbasids frequently had to contend with theological controversies, which were no less serious a challenge to their authority. But with the period of conquest at an end, the spirit of *jihad* had declined. Whereas the great conquests of the past had been achieved entirely by Arab forces, the Abbasid Caliphs became increasingly dependent on non-Arab troops to deal with rebellious provinces and religious revolts. This was a fatal error; in the time-honoured fashion of mercenary corps these Turkish, Berber and Khorasan troops began to assert themselves and usurp central authority.

North Africa became autonomous as early as 800, and Khorasan in 820. In 868 a Turkish soldier of fortune, Ahmad b. Tulun, managed to gain control of Egypt and Syria, which he and his descendants ruled until 905. These countries were later brought back into the Empire by the more vigorous Caliph Muktafi, but worse was to follow.

The schismatic Alids (followers of Ali, the Prophet's son-in-law and fourth Caliph) had long been a thorn in the side of orthodoxy and legitimacy, but in 909 a leader emerged who (with the aid of a Berber army) vaulted their cause to prominence by declaring a Fatimid Caliphate in North Africa. This was a direct challenge to the Abbasid claims to be the sole religious arbiters in the Islamic world, in fact, it meant that there were now three 'Commanders of the Faithful' (in Baghdad, Cordova and Mahdiya), each aspiring to this role. The Fatimids used North Africa as a springboard to conquer Egypt, in 969, and their foundation there of a new capital, al-Kahira (Cairo) is a landmark event in Islamic history; essentially, Egypt regained its ancient status as an independent state—and Fatimid power soon spread to Syria and Arabia. This meant that not only had the Abbasid Caliphate lost control of huge swathes of their territory, but that their spiritual, Sunnite legitimacy was now being seriously challenged by a proselytizing Shia movement with a secure base of its own. Even at home, in Baghdad, the Caliphs had been forced to yield to the Buyids, a group of Shi'ite adventurers. But the later 10th century was to prove to be the high-water mark of Shi'ism. Although the Abbasid Caliphs were now reduced to mere figureheads and much of the Muslim world was under the control of one or another Shia sect, there was to be a Sunni counter-offensive from an unexpected quarter.

1.7 The Sunni Revival

The theological quarrels between Sunni and Shia that raged throughout the Muslim world at this time (ostensibly about rights of spiritual succession) were usually far more complicated than they appeared on the surface. As always in these cases, religious differences often masked ethnic and political enmities. Moreover, the views of the rulers did not always reflect those of the general population. Despite the apparent ascendancy of Sh'ism, the early 11th century began to see a revival of Sunni fortunes. The disintegration of the Abbasid Empire had created opportunities for nomadic, recently Islamicised, Turkish tribes from the Asian steppes. Two of these in particular, the Seljuks and the Ghaznavids came to play leading roles in the Islamic world; in time they became powerful military aristocracies, just as the Arabs had in the past. Most importantly for Islam's religious constitution, they were both ardent champions of the Sunni cause. In fact, the Seljuks took the lead in consolidating Sunni Islam.

The broad movement that has become known as the 'Sunni Revival' was a gradual rather than sudden change of direction. The Seljuks restored the Abbasid Caliphate—from which they then drew their religious legitimacy, as did the Ghaznavids (the leaders of both of these tribal dynasties were thereafter designated as Sultans). By the second half of the 12th century, these new dispensations had established a new 'ecumenical' Sunnism. This new orthodoxy, which managed to include many previously contending factions, gained wide acceptance in the Islamic heartlands of Iraq and Persia. These changes were also felt in Egypt, where the schismatic rule of the Fatimids was ended by a Turkicised Kurdish dynasty, the Ayyubids, who also replaced Shi'ism with Sunnism as the official interpretation of Islam.

The break-up of the old Caliphate, and the disintegration of the Empire into jostling groups of smaller states, had seriously damaged the Islamic vision of itself as a coherent religio-political entity. The widely accepted settlement, both religious and political, of the Sunni Revival went a long way to restoring a sense of cultural confidence in the areas where it was accepted. This confidence is clearly reflected in the arts and architecture of the time. Among other artistic achievements, this was the period when the characteristic forms of the Islamic decorative canon reached their full development, and when Islamic architecture reached new heights of brilliance and innovation (see Chap. 4). Paradoxically, there was a price for this artistic and cultural revival. The new emphasis on a religious orthodoxy, however welcome it was to pious believers, involved a distinct narrowing of intellectual horizons and a retreat from the enthusiasm of earlier periods for pre-Islamic philosophical and scientific speculation (see Chaps. 2 and 3). Nevertheless, at the beginning of the 13th century the *Dar al-Islam* (The Islamic World) seemed to have attained a stability and confidence that it had not experienced for generations. This peace, however, was about to be shattered.

1.8 The Mongol Cataclysm

As we have seen, the original Islamic conquests were made by forceful Arab nomads, who launched themselves against the sedentary cultures of the Middle East. Essentially, what began as a series of raids (which were a long-established feature of nomadic life) was turned into full-blown conquest. The first Muslims came out of the desert, conquered vast reaches of the known world and managed to hold onto it. But these extraordinary successes inevitably led to the conquerors adopting many of the civilised values of the conquered lands—not least in defending their acquisitions from other marauders. In the process they became thoroughly civilised themselves.

This course of events was repeated by the 'Turkish Irruption' that began to plague the eastern end of the Islamic world at the end of the 10th century. These nomadic incursions, destructive as they initially were, eventually had the same sort of revitalising effect on the Dar al-Islam that the original Islamic forces had exerted on the old middle-eastern empires. There had been a long history of nomadic encroachment into Islamic territory; greater and smaller groups of barbarous tribesmen had been coming in, either as raiders or mercenaries, for centuries. Up to now Islam had managed to convert, absorb and civilise these outsiders. But the Mongol invasions that began in 1219 were of a different scale and order of ferocity than anything that the Dar al-Islam had ever experienced. From the beginning it was clear that these invaders were not interested in stealing from or joining civilised society in any capacity; their aim was to destroy it.

Genghis Khan became known as the 'World Shaker', which indeed he was. He and his successors, by the sheer numbers of troops they employed and by the tactics of systematic terror that they adopted, had a devastating impact on civilisations right across Asia. The Mongols first honed their ruthless skills in China, which they ravaged for five years. Peking was the first major city to suffer their appalling depredations. Merciless and bloodthirsty, they seemed to be consumed with a hatred of urban civilisation; cities were only fit for plunder, their inhabitants worthy only of contempt. Having devastated the great urban centres of northern China and massacred huge numbers of the population, the Mongols turned their attention towards the Khwarezm Empire, in the eastern-most part of the Islamic world.

They prepared their destructive campaigns very thoroughly. In 1219, the ancient oasis towns of Samarkand and Bukhara were attacked simultaneously and without warning; they were besieged, sacked, burned and their inhabitants slaughtered. From here, Genghis Khan turned his attention to Khorasan in eastern Persia, and embarked on one of the most appalling campaigns of blood-lust ever recorded. City after city was subjected to his methodical ferocity, population after population was annihilated; an entire cultural landscape was razed. It has been estimated that, in Central Asia and Khorasan alone the Mongols left fifteen million dead, and they were far from finished.

In 1257 Genghis Khan's grandson, Hulagu, returned to Persia and Iraq on a further campaign of terror and violence. This time the aim was to subdue 'all the lands of the West'. This naturally included the city of Baghdad, where the Abbasid Caliph was still the Commander of the Faithful (if in name only). The atrocious pattern that the Mongols had established in Central Asia and Khorasan was thus repeated in the Islamic heartland. In 1258 Baghdad was besieged by a huge Mongol army and pounded by their artillery for six days. When it was obvious that the city's cause was lost the Caliph, al-Mutasim, gave himself up to the invaders' mercies; he was executed by being rolled in a carpet and trampled to death by horses. The death of the last of the Caliphs was soon followed by the utter destruction of the greatest of all Islamic cities. Upwards of a million of Baghdad's inhabitants were killed; whatever could not be carried away was burned or smashed. The mosques and palaces, the accumulations of art and literature, the wealth of five centuries of Islamic culture went under in a terrible orgy of destruction. The Mongols marched off with their booty through streets that were choked with corpses. It was an incalculable loss, from which Baghdad, Iraq, and the greater Islamic world, never fully recovered.

Hulagu then pressed on to Muslim Syria. The city of Aleppo offered resistance but fell in 1260 and was subjected to the, by now routine, mass-slaughter and sacking. Damascus surrendered to the hordes and was humiliated, but spared the worst atrocities. Mamluk Cairo was next in line; it received the usual imperious summons to capitulate, but at a critical moment Hulagu received news from Mongolia that his brother Ogedai, the Mongol Great Khan, had died and he hastily returned home to take part in the election for succession. A Mongol force was left behind in Palestine, but the Mamluks defeated them and Egypt was saved. This was a turning point for Islam; it was the first time that the Mongols had suffered defeat on this scale, and they never returned. Because Egypt avoided Mongol depredations, and because of its victory over them, Cairo, to a great extent, took over Baghdad's role and became the repository of the remnants of the old Arabic civilisation.

1.9 The Islamic Revanche: The Ilkhanids and Timurids

The classical age of Islam (and Arab ascendancy) ended in 1258 with the fall of Baghdad, but it is a testimony to the inherent strength of Islamic culture that it was able to recover from the devastating onslaught of the Mongol invasions and rise to further glories. In fact, some of its great cities never recovered, but a process of recovery began when the descendants of Hulagu converted to Islam and founded the Il-Khanid dynasty. Once their rule was firmly established the Mongols assumed the role of a warrior aristocracy and, almost as a reaction to the destructiveness of the past, became great patrons of the arts and architecture. It took the better part of the 13th century, and a period of wholesale reconstruction, for a new Mongol-Islamic art to establish itself, but when it did the Islamic tradition was enriched by an extended repertoire of Far-Eastern influences. Il-Khanid architecture continued the Seljuk tradition but on a more massive scale. Unfortunately very little of either art or architecture from this period has survived.

The recovery of Eastern Islam (Iran, Khorezm, Afghanistan) was set back by a further, and final, nomad onslaught with the rise of Timur, who began his appalling career around 1360. Timur (Timur-i-leng, Timurlane) nominally a Muslim, was a savage destroyer of cities. His name became a synonym for terror, and his memory is forever associated with the vast heaps of skulls that were piled up beside the cities that he conquered. After Timur's death there was a period of political chaos, but his descendants, like those of the Mongols, were more inclined to repair, rather than extend, the ravages of their blood-thirsty forebear. In fact the Timurid dynasty oversaw an extraordinary revival of the Islamic faith and culture. The latter was partly made possible because even at the height of their ravages both the Mongols and Timur recognised the worth of craftsmen, they were often spared or taken as captives, so when peace retuned they were able to resume their trades.

The Timurids became lavish patrons of the arts and architecture. The vast Imperial projects that they initiated, which continued the Il-Khanid tastes for monumentality of design and magnificence of decoration, promoted all the crafts associated with building and were responsible for a renaissance of all the arts. The synthesis of Mongol-Turkic and Arabic styles created at this time brought entirely new and highly refined forms both in architecture itself, in architectural decoration and in the arts of the book, which they were also keen to promote. All of these developments were to have a lasting effect on the course of eastern Islamic art.

The Timurids moved their capital from Samarkand to Herat, but like so many before them, their brilliance as patrons was not matched by the sort of political and military skills that were necessary to stay in business at that time. During the 15th century their territories were constantly being encroached upon by aggressive Turcoman tribes. Eventually their Empire was reduced to Herat and the province of Khorasan, where the final flowering of Timurid culture took place.

1.10 The Early-Modern Islamic States

The Timurids were the last great Islamic dynasty of nomadic, steppe origin. The adoption of firearms and more advanced military techniques effectively put an end to further large-scale invasions from the Eurasian steppes. The next stages of Islamic expansion and consolidation were dominated by states that were ruled by Turkish dynasties. These included the *Mamluks*, rulers of Egypt and Syria; the *Ottomans*, whose empire eventually encompassed North Africa, Asia minor, most of the Near East and Arabia; the *Safavids* who ruled Persia and Afghanistan; and the Mughals who gradually conquered much of the Indian sub-continent. Each drew on the art and architecture of their immediate predecessors, but went on to develop their own styles; all departed from the modes of classical Islam in various distinctive ways (Fig. 1.2).

The Mamluks were originally the Turkish slaves of the Ayyubids. Although their rule was turbulent, with frequent turnovers of ruler, this period 1250–1514 was generally a prosperous one for Egypt and Syria, as a result of which culture and the arts thrived (see Fig. 1.1).

The Ottomans first emerged in the late 13th century as one of many small tribes of Turcoman tribesmen, and were loosely attached to the Seljuks. Very gradually they took over Asia Minor (including Constantinople in 1453, overran a large part of the Balkans, captured Egypt in 1514 and Hungary in 1526.

Their heyday followed the conquest of Constantinople, which became the centre of western Islamic civilisation.

The Safavids were Turkish-speaking, but after establishing monarchical rule in 1501 they imposed Shi'ism as the state religion and claimed semi-divine status as

From National Library of Cairo. Date not known except being during the Mamluk period (1250-1517). Part of the UNESCO's Memory of the World. For graphic, see (Tiling Search Web Site 2017, data183/CD2)

Fig. 1.1 Illuminated page from a Mamluk Qur'ān

reincarnations of the Shia Imams. These moves provoked perpetual hostility from their Sunni neighbours (particularly the Ottomans), but gave Persia a distinct Islamic and national identity that it has retained to this day.

The reign of Shah Abbas, 1588–1629, is recognised as the Safavid golden age, with the Shah building his magnificent new capital at Isfahan in 1596.

The Mogul Dynasty was founded by Babur, a Chaghatay Turk descended from both Genghis Khan and Timur. In 1504 he raided northern India, but the dynasty was only firmly established by his son Humayun in 1555. The most glorious period of Mughal rule followed with the Emperor Akbar who reigned for the fifty years from 1556. Akbar was followed by other capable rulers whose patronage allowed Islamic-Indian culture to develop distinctive styles in art and architecture. The Mughals gradually declined as a power, until the dynasty was finally deposed by the English in 1858.

Ahmad Ibn Tulun, Cairo, 876-879 (Patterns in Islamic Art web site 2017, EGY 0512). This design was used in a Roman mosaic in Ouzouer-sur-Trezee, France, but with different ornamentation (Tiling Search Web Site 2017, data15/S2)

Fig. 1.2 Classical design to Islamic use

1.11 Spain and the Maghreb

The far west of the Islamic world, which included the interacting cultures of Spain and Morocco, was never entirely unaware, or uninfluenced by the tumultuous events in eastern Islam, but its location tended to place it outside the main stream of Islamic history. This, however, did not prevent it from having a fairly turbulent history of its own.

The Spanish peninsula was invaded during the early Islamic conquests, when Arab and Berber forces crossed the Straits of Gibraltar in 711. The first important dynasty was established by the Umayyad prince Abd-ar-Rahman, who was a fugitive and the sole survivor of the Abbasid massacres of the Umayyad dynasty in Iraq. The western Umayyads made Cordova their capital, which by the 10th century was a flourishing centre of culture, trade and manufacture comparable in size and wealth only to Cairo and Baghdad. The dominant figure of this period, and the greatest ruler of the dynasty, was Abd-ar-Rahman III, who adopted the title of Caliph and Commander of the Faithful.

Despite its relative isolation from the rest of the Muslim world, and the difficulties in maintaining central rule in such a geographically diverse country, Islamic culture in Spain under the Umayyads reached great heights. It finally disintegrated in the early 11th century as a result of the usual combination of internal rivalries and external pressures. The Umayyads finally disappeared in 1031 after which there was a period of political fragmentation into innumerable factions (*taifa*). The capture of Toledo by the Christians in 1058 prompted an appeal to a North African warrior tribe, the

Almoravids, 1054–1147, who conquered the whole of Muslim Spain by 1090. The unity they imposed did not last. Another wave of invaders, from the Atlas mountains in Morocco, the Almohads, 1130–1269, added to the 'Africanisation' of Islamic Spain in this middle period.

After the demise of the Almohads three smaller principalities came to dominate the Maghreb in the late Middle Ages: the Marinids in Morocco 1196–1465; the Zayanids of western Algeria 1236–1555; and the Hafsids of eastern Algeria and Tunisia 1229–1574. From the 16th century most of the Maghreb fell under Ottoman rule.

In Spain, the late medieval history of Islam was marked by a serious diminution of its territories. Cordova and Seville were lost to the Christians in 1212, leaving Granada alone as an isolated emirate. Miraculously, the Nasrid dynasty managed to maintain an Islamic presence here for a further two and a half centuries, 1232–1492; miraculous too, is the fact that this small besieged outpost continued to be an important cultural and manufacturing centre, famous for its poets and scientists, its textiles and pottery, and of course for its marvellous royal palace, the Alhambra.

Granada finally fell to the Catholic alliance of Ferdinand and Isabella in 1492, the very year that Columbus discovered America. However, Islamic cultural attitudes and style tended to persist in most regions of 'reconquered' Spain. In fact the *Reconquista* waged by Christian Kings on Muslim Spain (*Al-Andalas*) was complex and protracted, taking place over seven centuries, with extended periods of respite between the warring parties. Initially this conflict was a typical medieval war of conquest and only gradually acquired the religious/ideological aspect by which it later came to be portrayed. Since they constituted such a large proportion of the conquered populations Muslims were permitted to remain in the territories that fell to the Christian forces and, for reasons of expedience, were allowed to practice their faith. These Mudéjares (from the Arabic *mudazzan*—remainers), continued for centuries to contribute their civilised skills to what became a flourishing, essentially hybrid, Mudéjar culture. Highly evolved crafts such as ceramics, woodworking, brickwork and plaster decoration continued to be practiced by Muslims, but were now mediated by the requirements of their new Christian patrons. This resulted in distinctive Mudéjar forms in art, music and architecture, which left an indelible impression on the Spanish national consciousness. Ironically, some of the best surviving examples of Moorish architecture in Spain are still to be found in the many palaces, churches and other monuments commissioned by the ruling Christians, but executed by Muslim, Mudéjar craftsmen.

1.12 Imperialist Encroachment and Islamic Decline

During the period described as the Middle Ages in Europe the Islamic world was at the height of its power, wealth and creativity. For much of this time European Christendom was on the defensive against Islamic, particularly Turkish, policies of expansion. The turning of the tide of Islamic pre-eminence as a world culture in favour of its quarrelling European rivals is marked by three important dates. Islam still mourns the loss of Spain in 1492, an event that effectively marked the end of the Medieval and the beginning of the Modern period for western Europe. This period also marked the beginning of European imperialism, much of which was to be at the expense of the Islamic word. Islamic expansion into Europe came to an abrupt end in 1683, with the failure of the Ottoman siege of Vienna. From this time Islamic forces suffered defeat after defeat when confronted by Europeans. The British gradually gained control of Mughal India; the Russians pushed ever further into North and Central Asia; the Portuguese went into Africa and Indonesia. In 1789 Napoleon occupied Egypt (and was soon supplanted by the British).

By the early 20th century most of the Islamic world had been incorporated into one or another of the European Empires (principally those of Britain, France, Russia and the Netherlands). The Turkish Ottoman Empire, which had been in decline for the past century, was defeated in 1918 and its territory partitioned between the victorious British and French Empires. In a move that had enormous implications for modern Islam, Turkey was liberated by secular nationalists in 1922, who then went on to abolish the Caliphate in 1924 as part of their policy of modernisation. The rest of the Islamic world fought its way clear of imperial domination in the years after the Second World War, but was then faced with seemingly intractable problems of adjusting to modernism, particularly that of forming a loyalty to a nation-state, an alien concept to traditional Islam. The lingering sense of religious unity, but one without even the vestige of an overarching spiritual authority, still presents the greatest difficulty to modern notions of an Islamic identity, and by extension, to a distinctive modern Muslim culture.

Chapter 2
The Scientific Contribution

> The Greek material received by the Arabs was not simply passed on by them to others who came after; it had a very real life and development in its Arabic surroundings.
>
> *De Lacy O'Leary, How Greek Science Passed to the Arabs.* (O'Leary 1949, p. 4)

2.1 The Greek Connection

From the time that Muslim forces overran the Byzantine provinces of Syria and Egypt in the mid 7th century it was inevitable that Islam would encounter the traditions of late-Classical science and philosophy that were still being taught in the Greek schools there. However, this process of cultural assimilation was slow to develop. The initial intellectual exchanges between the Muslims conquerors and the Christian majority were entirely dominated by religious disputation. Indeed, during the first Islamic dynasty, the Damascus-based Umayyads, there was little interest at all in the secular aspects of Classical culture.

This was the extraordinary period of Islamic imperial expansion, and at the time the Arabs were far more interested in learning what they could of Roman military skills—which of course they went on to apply so successfully. It was in the following dynasty, that of the Abbasids, during Islam's second century, that interest in Greek science and philosophy really took off (Fig. 2.1). The Abbasids were more interested in consolidation than expansion, of controlling the vast territories now under their dominion. They founded a new capital at Baghdad, which effectively, became the focus of an entire civilisation. Here, under the enlightened rule of Harun al-Rashid, there began a sustained programme of translation of all that remained of Greek science and literature.

Harun's Caliphate, which began in 786, is generally recognised as Islam's Golden Age. Scholars, engineers, scientists and craftsmen of all kinds flocked to Baghdad from every part of the Empire. Talent was recognised, and well rewarded. At this time there was a dawning realisation of the value and usefulness of Greek mathematics,

B. Wichmann and D. Wade, *Islamic Design: A Mathematical Approach*, Mathematics and the Built Environment 2,
https://doi.org/10.1007/978-3-319-69977-6_2

Fig. 2.1 Hispano-Moorish Astrolabe, Spain?, c. 1300. Contemporary with the Alhambra

medicine, astronomy, and of its scientific knowledge in general. This new enthusiasm, for what almost amounted to a second Revelation, was actively encouraged by Harun. Schools of translation were established, and original manuscripts were brought in from every available source.

During the reign of his son, Caliph al-Ma'mun 813, the work of translating Greek scientific and philosophical literature was formally institutionalised. A 'House of Wisdom' (Bayt al-Hikma), which comprised an academy and library, was set up in 830, and continued to translate, revise and teach the Classics for the better part of two centuries. The appreciation and appetite for Classical knowledge was such that, by the end of the 9th century, the entire scientific literature of the Greeks was available in good Arabic translations—and this despite the fact that many contained doctrines that were at variance with those of Islam.

There are various aspects of this Islamic collation of Greek knowledge that are worth noting. Firstly, the primary sources of the literature were the Christian Syriac schools, which meant that the major part of the material that Muslim scholars received was that which was still valued in the late Hellenistic period, they were unaware of Greek poetry, drama and the works of its historians and orators. Nevertheless, there was a great body of knowledge from this source, and this was supplemented by older Greek material that was brought in from centres in both Persia and India, where Greek mathematics and science had survived and had developed independently.

As a result, Baghdad became the first great centre of Islamic scientific learning and philosophy, and its scholars were soon making original contributions of their own in such areas as astronomy, optics, algebra and trigonometry. In particular, the central role of mathematics was recognised and widely practised. Later, this knowledge was disseminated to other important centres such as Aleppo, Damascus, Cairo, Cordova and Samarkand. Philosophy, on the Greek model, was also to enjoy an independent life in this new Islamic setting.

2.2 The Pythagorean/Platonic Tradition

By the time Muslim scholars encountered Greek thought, Christian Neoplatonic ideas had been a powerful influence for several centuries. This meant that most Islamic interpretations of Greek philosophy tended to be viewed through the lens, as it were, of these later doctrines. In fact, from an Islamic standpoint, the Christian affiliation with Neoplatonism served to 'disinfect' this philosophy from any taint of paganism, making it more acceptable. The question of reconciling Classical and late-Classical philosophical systems with the revelations of the Qur'ān was always a factor in the history of philosophy in Islam; doctrines that affirmed a divine unity, such as those of Plato and Aristotle were, naturally, more favourably received.

The Neoplatonists themselves traced their roots back to the semi-legendary Pythagoras (6th century B.C.) and the school that developed his ideas. The Pythagoreans were the first to believe that the structure of the universe was to be found in mathematics—'All things are made of numbers'—and it can be fairly said that they laid the foundations of both arithmetic and geometry in the Western tradition. This school was much concerned with ratios and proportions (they also uncovered the laws of musical harmony), and seem to have ascribed mystical properties to both numbers and geometrical figures. For the Pythagoreans, numbers and proportions took the place of the Gods. They had a separate existence of their own, entirely independent of men's minds, the contemplation of which was a form of devotion or prayer.

Plato was greatly influenced by these theories and adopted their belief that number and form were the keys to a deeper understanding of the universe. He was also sympathetic to their perception of the gross material world as a place of corruption and illusion. Plato's philosophical ideas are extensive and not easily summarised, but one consistent theme was that of a supersensible realm of 'Forms', of which the world of ordinary experience was an imperfect copy. He was deeply interested in geometry and clearly felt that its method, which produced clear and definite proofs, could be more generally applied. In the Platonic view the world of Forms or Ideas is separate and superior to our world of ordinary experience—and free of its illusions.

This proposition—the existence of a place, beyond our immediate sense-experience, of timeless perfection—colours the whole range of Plato's thought. He had a very low regard for the art of representation, seeing this as 'a copy of a copy', or 'a third removal from the truth'. For Plato the truly beautiful could not be conveyed by any work of representation or imagination; at best these could only ever be conditionally beautiful. True beauty had to express at least some of the eternal quality of his 'Forms', the terms of which he seems only to have found in geometry.

2.3 Philosophy in Islam

The Neoplatonists, who conveyed Plato's philosophical ideas to the Islamic world, had in fact elaborated his philosophical system into a complex cosmology of their own. This movement originated in Alexandria in the 3rd century (long after the decline of Classical Athens). It was eclectic and was influenced by Pythagoras, Aristotle and the Stoics as well as Plato. In its later development it absorbed Jewish and Christian precepts. The main aim of its founder, Plotinus (200–269), was to connect with the supreme unity, the source of all existence and all knowledge, through mystical, ecstatic union. In this system, the lower, material levels of existence are a sort of overflow of the divine fullness. These, and later Neoplatonic speculations, exerted a considerable influence on Islamic philosophy, and on Islamic mysticism (Sufism).

In time, as they became more discriminating, Muslim scholars were able to separate out the older Classical philosophies from later accretions, and to make their own interpretations of this original material. In the three or four centuries following the founding of Baghdad the Islamic world produced many outstanding philosophers, whose contributions were appreciated well beyond the Islamic sphere. Among the most important of these were al-Kindi (d.870), al-Farabi (870–950), Ibn Sina (Avicenna) (980–1037), and Ibn Rushd (Averroes) (1126–1198). Each of these was concerned with producing a version of Greek philosophy for Muslims, and each incorporated Platonic and Aristotelian concepts (especially the latter) in their philosophical systems. As indicated above, they were also strongly influenced by Neoplatonic cosmological ideas. All of these philosophers were translated into Latin in the Middle-Ages, introducing Classical thought to the Christian West where it acted as a stimulus to philosophical and theological speculation.

There were many other important individual philosophers in the Islamic world during the period of the Abbasid Caliphate, and various esoteric schools. In this Islamic setting, philosophical ideas were frequently bound up with those of religion and politics, which often meant that it was expedient for groups of like-minded scholars to come together in secretive associations. The 'Faithful Brethren of Basra' (Ikhwan as-Safa), which formed in the second half of the 10th century, is the best known of such groups. The Brethren were Encyclopaedists; they compiled a sort of summary of all the knowledge of their day, in some fifty volumes (the *Rasa'il*). Drawing on Pythagorean sources, they attempted to combine religion and philosophy in a unified world view. Following Pythagoras and Plato they saw numbers and proportions as the key to a deeper understanding of nature. They firmly believed in the moral value of this knowledge—and that it was an essential aspect of beauty.

It is unclear to what extent these ideas, or those of any other of the Islamic philosophers, might have influenced the course of Islamic art, but these concepts were definitely in the air at the time. The intellectual climate was pervaded with notions of Platonic idealism.

2.4 Religious Revival, Scientific Retreat

As we have seen, the history of Islamic thought was marked by the encounter between the theologies that developed from the Qur'ānic Revelation and the Hadith, and various pre-Islamic schools of philosophy (both Classical and late-Classical). In the end though, the latter was absorbed by the former; philosophy was reformulated to theological sensibilities, rather than the other way round. But the working out of this collision of belief systems, rather like the movement of tectonic plates in geology, was a gradual, grinding process.

Serious differences in the formulation of Islamic doctrine arose virtually from the time of the Prophet Muhammad's death but, under the rule of the early Caliphs, broad notions of orthodoxy developed, if for no other reason than to support the status quo. The more literalist defenders of the Faith were always suspicious of philosophy, but the cultural confidence of Baghdad's Golden Age period exposed Islamic thinkers to many new influences—and allowed them to take a more objective view of their own theological assertions. Under Caliph Ma'mun, who founded the 'House of Wisdom' and encouraged science, a rationalist interpretation of Islam (Mu'tazilite) became the official doctrine (in 827). Using Greek philosophy as a model, the Mu'tazilites employed argument, reason and dialectic to establish their credentials. The rationalising, even free-thinking, tendencies of the times were also evident in the works of the physician Ar-Razi (Rhazes), and the philosopher al-Farabi. Farabi actually proposed an allegorical interpretation of the Qur'ān, which provoked serious opposition from Traditionalist critics.

The Rationalist era of Islam however, was short-lived. *Mu'tazilite* doctrines fell out of favour and were finally suppressed under Caliph Mutawakkil in the mid-10th century. This was against the background of the increasing instability of the vast Abbasid Empire; by this time the political authority of the Caliphate itself was under serious threat. Soon after this the political unity of the Empire began to disintegrate into a series of breakaway states, Muslims came to depend on religious, rather than political, unity, which in turn led to a general narrowing of intellectual horizons.

The turmoil and insecurity of the 11th century (largely the result of increasing pressure from Turkish incursions) saw the emergence of an ecumenical, pietistic version of Islam. But this reaction was accompanied by an attitude of intolerance towards perceived heretical beliefs, and proved inimical to scientific and philosophical activities. By the 12th century the Neoplatonism expressed in the *Rasa'il* of the Faithful Brethren, and other philosophical literature, was regarded as irreligious, and publicly burnt. The final flickers of the Graeco-Islamic intellectual enlightenment were expunged in the colossal tragedy of the Fall of Baghdad to the Mongols (see Chap. 1).

2.5 Continuity and Transmutation

The enormous achievement of the Golden Age in collecting, translating, disseminating and building on, the philosophies and sciences of the Classical past was of immeasurable importance. Ironically, the Western infidel nations were eventually to be primary beneficiaries of this fund of knowledge; it was instrumental in pulling Europe out of its Dark Ages, and laid the foundations of the Enlightenment. The names of Islamic scholars feature on the first pages of most histories of European science, mathematics, medicine and astronomy—and of course much of the rich tradition of Classical philosophy was first conveyed to the West in Latin translations from Arabic sources.

However, the cultural attainments of the Hellenised civilisation that the early Muslims encountered, and that their successors so readily adopted, did have a lasting effect on Islamic civilisation; in fact it helped to create it. Many of the norms of Islamic life were formed as a result of this contact. In essence there was a continuity of late-Hellenistic cultural values into the Islamic sphere. The clearest outward expression of this legacy may be found in Islamic art.

The most familiar aspects of this art—its preoccupation with symmetry, proportion and spacing; its highly geometrised aesthetic; its other-worldly, 'eternal' qualities—are essentially an expression of Platonic, or rather Pythagorean/Platonic, philosophical concepts. It is unclear just how this transmutation, from the realm of ideas to the realm of art, was effected. It is the case, however, that at the very time that *falsafa* (philosophy) was coming under increasing pressure from religious orthodoxy, Islamic aesthetic sensibilities appear to crystallise around essentially Platonic notions of beauty. In fact (again ironically), the geometric and arabesque decorative modes, which are now so completely associated with Islam in all its manifestations, were originally adopted as the identifiable style of a renascent, Sunni orthodoxy.

Classical philosophy, always treated with suspicion by the narrowly religious, could not thrive in the spiritual and political turmoil that characterised the Islamic world from the 12th century on. But pure geometry could never be considered as heretical, and the interplay of Platonic figures on the Euclidean plane clearly did not violate any injunction in the Qur'ān or the Hadith. But the connection with the Classical past was never entirely forgotten. In a revealing passage in the Introduction to his 'History of the World', the famous 14th-century author Ibn Khaldun makes various observations, presumably of fairly widespread currency, about the craft of carpentry:

In view of its origin, carpentry needs a good deal of geometry of all kinds. It requires either a general or specialised knowledge of proportion and measurement, in order to bring forms from potentiality into actuality in the proper manner, and for the knowledge of proportions one must have recourse to the geometrician. Therefore, the leading Greek geometricians were all master carpenters. Euclid, the author of the *Book of Principles*, on geometry, was a carpenter, and was known as such. The same was the case with Apollonius, the author of the book on *Conic Sections*, and Menelaus, and others.

The Craft of Carpentry from the *Muqaddimah* (Khaldun 1967, p. 322).

Chapter 3
The Religious Dimension

> You shall not make unto you any graven image, or any likeness of anything that is in heaven above, or that is in the earth beneath, or that is in the water under the earth; you shall not bow down yourself to them, nor serve them.
>
> *The Second Commandment; Exodus 20:4-6*

3.1 'There Is No God but God'

From its very inception the Islamic mission was dedicated to a revival and purification of the 'religion of Abraham', that is to say, the Judeo-Christian tradition. It adopted the strict, monotheistic credo of the Old Testament, and was hostile to the paganism that was still flourishing in Arabia. Muhammad's first act, following his acceptance as a religious leader by the townspeople of Mecca, was to destroy the 360 idols around the Ka'ba. Up to this time Mecca had been an important centre of pagan pilgrimage.

The Ka'ba

B. Wichmann and D. Wade, *Islamic Design: A Mathematical Approach*, Mathematics and the Built Environment 2,
https://doi.org/10.1007/978-3-319-69977-6_3

Like so much else in Islam, this excoriation of idolatrous practices was a clear continuation of the Biblical prophetic tradition—but Islam brought a new understanding of its basic religious principles. The Islamic rejection of any form of idolatry is bound up with its rejection of any notion of divine intercession. The individual has a direct line to God; ultimately, redemption and damnation are individual matters. The Islamic attitudes towards images and imagery (a subject to which we shall return) have always been concerned with the purity of the human/divine relationship. The worship of idolatrous images, in this view, is therefore both delusory and useless. For other imagery, interpretations vary, but in general, representations that do not seek to create an illusion, or pretence of reality, are acceptable if kept away from any place of prayer.

3.2 The Qur'ān

The Qur'ān, by comparison with the Jewish Book of the Law, has very little to say about images. The Torah makes numerous and quite explicit prohibitions against figurative imagery of any kind. By contrast, the few references to idolatry in the Qur'ān are not directed to figuration as such. The Jewish horror of divine representation extends even to the accidental creation of images, and the inadvertent bowing before false gods. This extreme image-phobia led to a culture in which art had a greatly reduced role. This is very obviously not the case in Islam. Islamic concepts, whilst regarding pagan idolatry as an abomination, rarely expressed the same abhorrence of non-religious imagery. In fact, prohibition of iconography was almost unnecessary in this cultural setting—the confession of Muslim faith includes the declaration that 'There is none like unto Him', reflecting the Islamic concept of an absolute and eternal Creator who is, by implication, utterly beyond representation. In other words, iconoclasm is bred into the very bones of Islam. Non-Muslims have always made much of the supposed Islamic ban on images as a prime determining factor of Islamic art. In reality, representational art was never suppressed, except in the strictly religious field—in any case such a purely negative influence could not in itself account for the many positive aspects of Islamic aestheticism.

The Qur'ān constantly reiterates the transcendent and inaccessible nature of Allah, with the implication that the Divine Nature can only be experienced through the Divine Word. For this reason there is no equivalent in mosques of the sort of iconography found in Christian churches, but its place is often taken by quotations from the Qur'ān, which can be quite extensive, and which gave rise to the rich traditions of Islamic calligraphy as an elevated form of architectural decoration.

Various other Qur'ānic themes have contributed to Islamic artistic sensibilities, including the *perfection* of the Creation ('Thou seest not in the Creation of the All-Merciful any imperfection', Qur'ān 67:3), the divine quality of *light* ('Allah is the light of heavens and the earth', Qur'ān 35:24), and a pervasive sense of otherworldly esotericism ('With Him are the keys of the secret things; none knoweth them but He', Qur'ān 6:59). Each of these qualities contribute to the well-known Islamic preoccupation with the dissolution of matter, for the hints of infinity in the patterns of which it is so fond and for the sense of space that it so often seeks. The aim seems always to escape imprisonment within earthly forms. The unwillingness to accept the 'counterfeits' of representation also account for this culture's preferences for abstraction in its decorative schemes.

3.3 Traditions (Hadith)

The Revelations that comprise the Qur'ān are, of course, the foundation of Islam, but the recorded sayings and deeds of the Prophet, the Hadith, constitute another body of religious texts. The Hadith do not have the scriptural authority of the Qur'ān, indeed they have varying levels of authenticity, but they are an important secondary source of religious guidance. There are various Hadith that refer to artistic creation. The most notable of these records Muhammad as saying that 'No angel will enter a house in which there are images'. Although this has been interpreted as an injunction against figural representation by more puritanical elements, it was almost certainly directed against household idols. Another Hadith warns the maker of images that on the Day of Judgement he will be required to breathe life into his creations, and on failing will be condemned, but this too was primarily concerned with idolatry.

Although the reservations expressed in these Hadith about the portrayal of living things obviously derive from the Jewish tradition, the Islamic interpretation makes a new distinction between vegetable and animal subjects, with the former being allowed, even in Mosques, and the latter banned. There is a similar line of demarcation between flat and sculpted surfaces, expressed in a Hadith that disapproves of art forms that 'create a shadow'; this is closer to the Jewish original. These concerns are clearly directed towards representations that try to create an illusion of animate reality, an enterprise that might be interpreted as an attempt to imitate Allah, who alone can create life. It is the case, however, that images of plants were used as decoration from the very earliest period of mosque-building.

3.4 Islamic Identity and the Image Controversy

The Islamic stance on the use of images for religious purposes was formed at a time when the Christian Church was involved in its own controversy about the subject. Christianity, like Islam, had originally followed Jewish precepts on the matter of images. In the course of time, however, this influence had receded, and the various Christian communities and sects had came to adopt very different views on the subject. In fact there were serious arguments within the broader Church for centuries as to whether or not it was proper, to use images of Christ, the saints and the martyrs as an aid to worship. Naturally, as in all theological matters, the partisans on either side were equally fervent in their beliefs. Islam encountered this controversy, and was obliged to intervene, when it conquered the Christian Middle East.

From an Islamic standpoint, the Christian use of images in their liturgy, together with the dogma of the Holy Trinity and various other articles of their faith, simply proved the extent of Christian departure from pure monotheism. To Muslims they were all anathema. In 721 the Umayyad Caliph Yazid II ordered all images (including those in mosaic) to be removed from the Churches within his domains, and all coins bearing figures of the Christian Emperor to be replaced with de-Christianised versions. In the process, Islamic primacy (and self-righteousness) was asserted and Christian faith and practice was on the back foot.

The effects of this Islamic iconoclast campaign against the Church, both within the conquered territories and in the heartland of the Byzantine Empire, were to be very far-reaching. The loss of vast swathes of their territory to the Muslims had already severely shaken Byzantium. There were many who blamed the Church's wealth and arrogance for invoking this divine punishment, and the long-running dispute concerning the use of images became the focus of a power struggle, and a virtual civil war, within the remaining part of the Empire. The Iconoclast Controversy raged on for well over a century, involving the Emperors themselves, some of whom were partisans for, others against, icon worship. Much blood was spilt in this crisis, which saw a constant fluctuation of fortunes between the opposing factions.

In the end the Iconophiles won the day in Byzantium. Icons were restored to the Churches in 843 (an event that is still celebrated in the Orthodox Church), and their use has continued right up to the present. For the Muslims, of course, the whole episode provided further confirmation (if any were needed) of the grievous errors accompanying Christian belief.

3.5 Mizan, Symmetry and Cosmic Equilibrium

Perhaps the most marked and abiding feature of Islam aesthetics lies in its dedication to symmetry principles. The arts of most cultures use symmetries to a varying degree, for various effects—but none place quite so much emphasis on the constraints of symmetry, or use it so consistently across the entire range of their artistic productions. Clearly, this aesthetic response derives from some deeply-held intuition on the underlying principles of existence.

Classical Greek ideas have undoubtedly contributed to this perception (see Chap. 2), but the motifs of cosmic equilibrium and the general harmony of the Creation also occur in several passages of the Qur'ān: 'The sun and the moon to a reckoning, and the stars and the trees bow themselves; and Heaven—He raised it up and set the balance' (Qur'ān 55:5). This notion of a cosmic symmetry, symbolised by the balance-scales (*mizan*), extends to ideas of eschatological justice and retribution, a most important feature of Islamic religious belief. On the Day of Judgement good works and misdeeds will be weighed to 'an atom's weight' (Qur'ān 99).

The *Mizan* is an important and well-known symbol in Islamic theology—some early literalists took the Qur'ānic reference to be an actual celestial entity. Interestingly, the term was also used to denote a ground-plan in architecture, and as a technical expression of musical patterns and rhythmic modes. In fact the introduction of symmetry concepts into cosmological speculation has very deep pre-Islamic roots; the scales were an important religious symbol in both Pharaonic Egypt and Zoroastrianism.

3.6 The Islamic Paradise (al-Djanna)

The Qur'ān refers to Paradise in several *suras* (chapters). The picture that is painted is of a luxuriant, sheltered garden full of sensual delights, all of which are essentially beyond human imagination: 'Gardens of Eden that the All-merciful promised His servants in the Unseen' (Qur'ān 19:61). In this exalted place the righteous experience the pleasures of the intellect as well as those of the senses. Not surprisingly, the promise of Paradise became a prominent theme in Islamic preaching.

It is also a major influence in Islamic art. In Islam the gratification of the senses and the search for perfection are not incompatible with spirituality. On the contrary, earthly beauty is seen as a pale reflection of the transcendent beauty of the Unseen. In this view it naturally follows that magnificent architecture and exquisite objects can put the believer in mind of the heavenly paradise to come (Fig. 3.1).

There are many direct references to Paradise in Islamic art and architecture, in mihrabs and prayer mats for instance, the image of the Garden is frequently

This is part of the enormous decorated walls of the 8th-century Umayyad palace of Qasr Mshatta, one of several of that dynasty's Desert Castles; located in present-day Jordan

Fig. 3.1 Islamic imagination

Fig. 3.2 Mosaic decoration from The Great Mosque, Damascus. This also shows an unbroken continuation of late-classical, Byzantine tradition of sacred architectural decoration, but the subject-matter, of a Paradise landscape, is decidedly Islamic (Patterns in Islamic Art web site 2017, SYR 0135)

symbolised, and carpets designs are often reminiscent of an idealised garden. Similarly, the use of floral and vegetal decorative motifs in an Islamic context are bound to refer to *al-Djanna*, however faintly. Beyond these more direct references however, the transcendent qualities of the delights of Paradise have exerted a broader influence across many modes of Islamic artistic expression (Fig. 3.2).

The high quality of Muslim workmanship in so many fields—in carpets, ceramics, woodwork, the arts of the book and in all the crafts associated with architecture—are well-known, as are its refined standards of taste. The aim seems always to be towards perfection, of both style and execution. At their best, the arts of Islam manage to combine the sensual with the spiritual. In this culture, Beauty, represented by colours and forms, and Perfection, as expressed in the production of artefacts, always seem ultimately to refer to the numinousness of the divine.

Chapter 4
The Evolution of Style

> One would enquire in vain for the masters who brought this system to its flowering or those who later opened up new ways for its development. This art is totally anonymous and it would contradict the artist's noblest charge, which was the liberation of the spirit from the transitoriness of worldly ties.
>
> *Ernst Kühnel, The Arabesque* (Kühnel 1977, P. 13)

4.1 Consistency and Variety

Islamic art has a recognisable aesthetic signature that somehow manages to express itself across an entire range of productions. The 'language' of this art, once established, was readily assimilated by each of the different nations and ethnicities that were brought within the Islamic sphere. Assimilated and built upon, because every region, at every period, produced its own versions of this super-national style.

This extraordinary consistency of styles and artistic preferences in the Islamic world clearly derive from a deeper, social consistency. Muslims have always held to the same basic system of belief, with its customary forms of religious observation, and all, despite national and ethnic differences, have identified themselves as Muslim first and foremost. Historically, this strong sense of identity and continuity has tended towards a high degree of social, and artistic, conservatism. As a result many forms and artistic concepts remained unchanged over the centuries. on the other hand, Islamic art has constantly demonstrated its capacity for the creative reinterpretation of accepted forms.

Much of the art of Islam, whether in architecture, ceramics, textiles or books, is the art of decoration—which is to say, of transformation. The aim, however, is never merely to ornament, but rather to transfigure. Essentially, this is a reflection of the Islamic preoccupation with the transitory nature of being. Substantial structures and objects are made to appear less substantial, materials are de-materialised. The vast edifices of mosques are transformed into lightness and pattern; the decorated pages

B. Wichmann and D. Wade, *Islamic Design: A Mathematical Approach*, Mathematics and the Built Environment 2,
https://doi.org/10.1007/978-3-319-69977-6_4

of a Qur'ān can become windows onto the infinite. Perhaps most importantly, the Word, expressed in endless calligraphic variations, always conveys the impression that it is more enduring than the objects on which it is inscribed.

4.2 Symmetry and Geometry

Another familiar characteristic of this art, which must also express something fundamental to the Islamic spirit, is its predilection for orderly, symmetrical arrangements in general and for purely geometrical decorative forms in particular. The influence of both philosophical and religious ideas on this aspect of Islamic art were examined earlier (in Chaps. 1 and 3 respectively), but the abstract, ideational nature of symmetry and geometry clearly fit with the Islamic taste for idealistic otherworldliness (Fig. 4.1).

From a purely doctrinal viewpoint, geometrical designs, being free of any symbolic meaning (which is the case in Islamic art), could convey a general aura of spirituality without offending religious sensibilities. In addition, the purity and orderliness of patterns and symmetries are able to evoke a sense of transcendent beauty which, at best, would free and stimulate the intellect (rather than trap it in the illusions of mere representation).

There is a certain disregard for scale in Islamic art that derives from this perception. Similar kinds of patterning, for instance, might be found on a huge tile panel or on a bijou ornament. This is because decorative effects, in an Islamic context, are never mere embellishments, but always refer to other, idealised states of being. In this view,

This pattern is common and widespread, it is found on the side of many minbars in Cairo in particular (Tiling Search Web Site 2017, data162/P077). The earliest example appears to be in Syria dated to 1163; the model for this reproduction is from 1335. This design features in (Bourgoin 1879, Plate 77)

Fig. 4.1 Minbar of the al-Nasir Mosque, Cairo

Detail of carved wooden door panel. Iraq, ca, 9th century CE

Louvre

Dish With epigraphic decoration. The Kufic inscription reads: 'Science has first a bitter taste, but at the end it tastes sweeter than honey'

Louvre

Decorative panel, geometric and Arabesque. Samarra, 9th century CE

Museum für Islamische Kunst, Berlin

Stucco & Plasterwork Ibn Tulun Mosque, Cairo (Patterns in Islamic Art web site 2017, EGY 0513)

Floriated Arabesque, Al-Salih Tala'i Mosque, Cairo. 555AH, 1160 CE (Patterns in Islamic Art web site 2017, EGY 0518)

Fig. 4.2 Examples of early Islamic art

scale is almost irrelevant. For similar reasons Islamic *ornemanistes* usually opted for acentric arrangements in patterning, avoiding obvious focal points—a preference that resonates with the Islamic perception of the Absolute as an influence that is not 'centred' in a divine manifestation (as in Christianity), but whose presence is an even and pervasive force throughout the Creation. A further analogy can be drawn between the patiently created repeats of the 'infinite' pattern (in all its varieties), and the familiar and unvarying customs of Muslim religious observances. In an Islamic context repetition is not tedious; on the contrary, it connects to the world of the spirit.

Whether in a religious setting or not, the work of Muslim artist/craftsmen always manages to convey a certain integrity, even nobility. In fact the distinction between art and craft is largely irrelevant in the Islamic world, but even when their works demonstrated surpassing skill and inspiration, the practitioners tended to remain anonymous. This is not surprising; in a culture whose ideal was submission to the will of Allah, it was quite natural to submit creative individuality to a perceived higher notion of beauty (Fig. 4.2).

4.3 Earlier Islamic Art

It was seen earlier (in Chap. 1) that the Umayyad's were the first dynasty to rule the newly-established Islamic Empire (from the mid 7th–8th centuries). During this early period, which was marked by conquest and consolidation, such art and architecture that was commissioned drew on pre-Islamic traditions (primarily those of Christian Byzantine and Sassanian Persia). Even so, there are clear signs of the emergence of Islamic aesthetic priorities. The main elements of decoration at this time derive from late-classical traditions of stone-carving, floor and wall mosaic and wall painting, but plaster decoration, introduced from the Hellenised East, is also used.

4.4 Baghdad, Samarra and Political Fragmentation; 8th–10th Centuries

The replacement of the Umayyad dynasty by the Abbasid (in 750) saw the removal of the capital eastwards to Baghdad, and later to Samarra. These two great cities were enormously important for the development of Islamic art and culture but, sadly, little has survived of either from the early period. From the fragments of architectural decoration that have been recovered there appears to have been a steady move away from naturalistic treatments, towards more abstract and repetitive forms. The long-established Eastern traditions of plasterwork and brickwork were adopted as the principal architectural materials in the Islamic East, and were to remain so for the following centuries (Fig. 4.3).

This window-grille from The Great Mosque, Damascus shows a direct continuation of the Syro-Roman tradition (Patterns in Islamic Art web site 2017, SYR 0118)

Fig. 4.3 Early window design

By the early 9th century, Turks from Central Asia become an ever more dominant political force, and from this time on their tastes exerted a strong influence on Islamic art generally. At the political level, these incursions meant that the Islamic Empire began to fragment into rival factions, with local dynasties establishing themselves in various regions. Islamic art in these turbulent 9th–10th centuries is represented by local styles, usually based on older precedents, but always bearing the imprint of Islamic tastes and limitations.

4.5 Stylistic Maturity; 11th–12th Centuries

As indicated above, the Abbasid Caliphate began to lose political power, and its Empire to fragment, at the turn of the 10th century. The most serious consequences of this political turmoil, for Islamic culture as a whole, were the accompanying religious controversies and multiplicity of doctrines. In time, in a reverse of the status quo of earlier centuries, Shi'ism became the dominant creed among the various secessionist dynasties, and for a period this interpretation looked set to dominate the Islamic world. But early in the 11th century there began a Sunni revival, which had both religious and cultural aspects. This movement, which saw itself as a restoration of traditionalism, was accompanied by an artistic revival that established many of the enduring forms of Islamic art and architecture—in particular, its canon of decorative art.

The Sunni revival, which began in Baghdad, gradually spread through the Islamic world. As it did so, it became associated with a range of new artistic and architectural forms, which became an identifying mark, a sort of symbolic language. These new stylistic terms were adopted by the Ghaznavids in Khorasan, the Seljuks in Iran, the

Zangids in Northern Syria, and the Ayyubids in Egypt. The 'classic' style of Islamic ornament which used distinctive epigraphic, geometric and abstract vegetal elements first came to maturity during this period. This decorative canon was eventually taken up in every part of the Islamic world (for more on the Sunni Revival see Necipoğlu (1995) and Tabbaa (2001)).

4.6 The Islamic Decorative Canon

The three elements of the Islamic decorative canon began to appear as early as the Umayyad period, but they crystallised into their classic forms during the 'Sunni Revival' (Fig. 4.4).

Calligraphy gives a visible form to the revealed word of the Qur'ān and is therefore considered the most noble of the arts (Moustafa and Sperl, 2014). It manages to combine a geometric discipline with a dynamic rhythm. Interestingly, none of its many styles, created in different places at different periods, has ever completely fallen into disuse. In the Islamic world it takes the place of iconography, being widely used in the decorative schemes of buildings.

Geometric patterns have always had a particular appeal for Muslim designers and craftsmen. They convey a certain aura of spirituality, or at least otherworldliness, without relating to any specific doctrine. In an Islamic context they are also quite free of any symbolic meaning. Above all they provide craftsmen with the opportunity to demonstrate his skill and subtlety of workmanship, and often to dazzle and intrigue with its sheer complexity.

Vegetal 'Arabesque' compositions are as ubiquitous in Islamic decoration as geometric patterns. It is difficult, without other indications, to determine where or when a particular composition of this genre might have originated. Like geometrical designs, these too are found across the entire range of media from book illustration to plasterwork, in ceramics, woodwork, metalwork and ivory-carving, even in carpets and textiles.

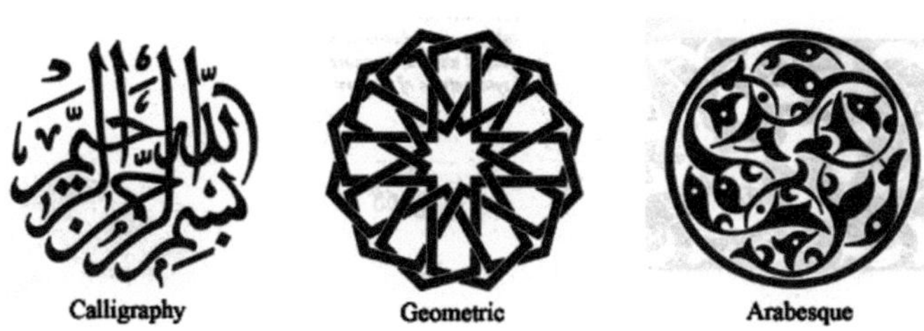

Fig. 4.4 Islamic decorative canon

4.7 The Craft Group

As indicated earlier, there was no distinction between 'art' and 'craft' in the medieval Islamic world, just as there was no sharp division between the notions of beauty and utility. The idea of the brilliant, lone individualistic artist was also absent. These are modern concepts. In fact, architectural and artistic productions generally tended to be the work of groups of anonymous craftsmen, whose occupations were usually hereditary and based within guilds or similar craft groups. However, given that the subject area of 'the decorative arts of Islam' has such an extensive geographical and historical reach, and involved a great variety of craft skills, generalisations of any kind about this broad subject may be misleading. Among such a diverse range of artistic traditions there were bound to be huge variations in working practices, social status, and indeed of expressive intent, in so many and varied creative processes.

As we have indicated, by comparison with life in the modern world, medieval Islamic society was highly conservative. Accordingly, working practices, just like the designs and motifs that were used, often persisted for generations with little change. There was no formal training; skills were acquired in the workshops, and were often handed down from father to son. The apprentice/master relationship was usually regularised however, and there is evidence that craft groups were formed into guilds (which had similarities with the trade guilds of medieval Europe). It also seems to have been the case that some of these guilds (in some periods at least) were affiliated to religious groups. Details of this kind are naturally difficult to uncover, as they tended to be bound up with the mystique and protectiveness of what were essentially closed professions. It is of more than passing interest though, that these religious connections were usually associated with Sufi fraternities, whose mystical outlook had been imbued with Neoplatonic concepts.

The production of specialised goods, and the particular skills involved, were often localised, but these skills often took artisans far away from their home. Occasionally these movements were less than voluntary. In the turmoil of war craft skills were generally prized as a form of booty and artisans could be carted off to the victor's base, sometimes en masse. Since the ruling power was usually the greatest patron of arts and architecture, it often happened that new artistic movements were initiated in this way. The dynasty founded by Timurlane was perhaps the most famous example of this effect. It occasionally happened that skilled craftsmen were welcomed as refugees after fleeing invasions and wars. The relative security of Fatimid Egypt attracted many Middle Eastern artisans during the turbulent 11th and 12th centuries, much to the cultural benefit of Cairo.

The transfer of skills and knowledge as a result of migrations and conquest is of course a time-honoured process. Huge numbers of skilled artisans from various traditions were brought into the Islamic fold as a result of the early conquests, and in the course of time these skills became part of the background of Islamic culture. But there was one craft technology in particular that was acquired in this way that had a profound effect on the course of Islamic civilisation on many levels (including the visual arts). This was the manufacture of paper.

4.8 The Role of Paper in Islamic Art and Architecture

Paper was a Chinese invention that had spread to Central Asia and Khorasan prior to the Islamic conquests of those regions. As soon as Islamic rule had been established in this eastern end of the Empire the usefulness of this product for bureaucratic and military purposes must have become obvious. The precise history of the uptake of this invention is somewhat hazy, but what is quite certain is that the relative cheapness and ease of production of paper ensured its rapid spread throughout the Islamic world during the 9th and 10th centuries—and that it transformed societies wherever it appeared. Paper was soon pressed into service by the Baghdad Caliphate, where it greatly facilitated their centralised rule, and Baghdad itself became one of many important centres of paper-making.

As paper became available throughout the Islamic world it was gradually adopted by artists and craftsmen. This usage began, naturally enough, among visual artists of higher status, calligraphers and those involved in the 'arts of the book'. The use of paper was slower to be adopted among craftsmen at a lower social position (woodworkers, stonemasons, weavers etc.,), partly because of the conservatism of their working practices, but also for cost reasons. As it became cheaper and more widely available however, paper became a standard tool in most crafts, and in the process it effected a transformation of the visual arts of Islam.

In essence, the adoption of paper into craft-based work practices introduced a certain separation of the design process from that of manufacture. This was bound to lead to a greater degree of organisation in design right across the board, from architecture through to the humbler crafts of pottery and weaving. Later this separation of design and making gave rise to a stylistic 'centralisation', such as that exercised by the many royal studios and workshops that were maintained by the royal courts. Many dynastic styles were established and maintained in this way.

Paper was a common commodity in the Islamic world as early as the 12th century and was already being used by various (usually high-status) artisans to produce preliminary drawings for their work. In the centuries that followed, the use of paper designs was gradually adopted as normal practice. This development undoubtedly fostered stylistic consistency across different media and promoted a common currency of style. There are many examples of close similarities between decorative motifs in architecture, book illumination and craft objects.

There is some evidence that the sort of complex patterning and arabesques that are such a characteristic feature of Islamic art first appeared as decorative elements of books (as did calligraphy, of course). As the 'arts of the book' enjoyed the highest status among crafts, it seems very likely that these designs would have gradually filtered down to other media. It is also clear that once pattern books and albums of design motifs came into being their repertoire could be applied to a whole range of artefacts, in different settings and at different scales. Designs could be copied, improved upon and, most importantly, they could travel (Fig. 4.5).

Roman mosaic, Tipasa, Algeria. Several other patterns with *khatems*, but all have extensive ornamentation with *guilloche* (p. 211). (A khatem is an 8-pointed star with a vertex angle of 90° (Tiling Search Web Site 2017, data191/ROMAN27) — an important constituent of patterns in Part II)

Fig. 4.5 Roman mosaic with khatem

It would certainly appear that there was, for centuries, a broad awareness and appreciation of the aesthetic aspects of ornament and a lively interaction between different crafts. The stylistic homogeneity of Islamic art referred to at the beginning of this note may have its foundations in Islamic cultural homogeneity, but its primary medium of transmission was paper (Bloom 2001).

Chapter 5
Materials and Media

As indicated in Chap. 4, the decorative arts in Islam are marked by a remarkable degree of stylistic consistency which has been applied to a broad range of materials, each of which had a craft-base of its own, the history of which often traces back to pre-Islamic times. The stylistic coherence within the Islamic world and the many variations of its basic themes across time, influenced as they were by local artistic traditions, are all part of the fascination of this art. In Muslim architecture virtually any surface may be regarded as worthy of receiving elaborate decoration. This is particularly apparent in religious architecture (Fig. 5.1), but this principle extends to woodwork, ceramics, textiles, metalwork, books and many other art forms.

5.1 Stonework

There is an East-West divide in basic Islamic architectural techniques that derives from earlier, pre-Islamic traditions. In the Persian/Iranian sphere of influence the principle building material tends to be brick, whereas in Egypt, Syria and Asia Minor, stone is far more common, at least for monumental building. The earliest Islamic monuments, dating from the Umayyad period, clearly continue the Roman/Byzantine tradition in their typical structures and in their use of dressed and carved stone. These techniques continued under later dynasties in Egypt and Syria (the Fatimids, Zangids, Ayyubids and Mamluks). This architecture is characterised by its monumental scale, its relative simplicity of form and a somewhat sombre tone. The deep carving, ordered in panelled schemes, with calligraphic bands and geometrical and arabesque motifs, make an impressive contrast against great expanses of undecorated surfaces. In Asia Minor the Seljuks, and later the Ottomans, continued the traditions both of ashlar building and of stone carving. Under the Seljuks a more plastic style of stone-carving was introduced, based on the stucco work of their predecessors in Iran. This gave

B. Wichmann and D. Wade, *Islamic Design: A Mathematical Approach*, Mathematics and the Built Environment 2,
https://doi.org/10.1007/978-3-319-69977-6_5

Fig. 5.1 Magnificent stone floor, The Great Mosque, Damascus. The date of this one is uncertain, but there are two in Cairo dated 1503-5 and 15th Century (2017, data182/BVIT)

Fig. 5.2 Example of stonework. Stonework from al-Azhar Mosque, Cairo, see (Patterns in Islamic Art web site 2017, EGY 1005). This example is dated to 1474. Two other examples are known from Cairo. Note that all the lines follow around a small path with an up/down alternation, see Sect. 11.5

rise to a rich tradition of that used all the familiar elements of Islamic decoration in a dazzling profusion of examples. The strength and vitality of this tradition of carved stonework continued up to the beginning of the 16th century, gradually becoming less exuberant during the Ottoman period. A separate tradition of architectural stonework developed on the Indian sub-continent, again derived from pre-Islamic sources. The early Turkish conquerors of India introduced entirely new, and in many ways quite opposite, architectural concepts to the sub-continent, but the synthesis that arose from the meeting of Hindu and Islamic traditions produced a great range of marvellous buildings, and countless examples of exquisite carved and inlaid stone decoration.

For an example of stonework, see Fig. 5.2.

5.2 Brickwork

As mentioned above, brickwork was the favoured building technique in the eastern Islamic provinces of Iraq and Iran, the tradition originating in the ancient civilisations of this area. Typically, however, in the hands of Muslim builders, brickwork was soon being used in quite novel and more decorative ways than in the past. In fact there is a

Fig. 5.3 Example of brickwork. Royal Mosque, Qazvin, Iran. The oldest part of the mosque is said to have been constructed at the behest of Harun al-Rashid in 807 (Patterns in Islamic Art web site 2017, IRA 2721)

well-defined progression in the use of brick in eastern Islam, from purely structural purposes towards ever greater decorative complexity. The first stage (in the 11th century) saw an increasing variety of brick bonds that created relief patterns of light and shadow to great effect. The enthusiasm for this technique was such that some buildings featured dozens of different bond-patterns, becoming veritable showcases of brick design. Later, carved ornamental inserts were used to break up the tedium of plain bonding; these were soon moulded before being fired, in a whole variety of motifs. In the next stage these brick inserts were glazed, a technique that lead naturally on to entire walls and domes being invested with coloured glazed bricks, by which time the structural and decorative functions of brickwork had been more or less separated out. Over time these glazed bricks were gradually reduced in thickness until they were virtually tiles—and the possibility of a whole new era of architectural ornament was created.

For an example of brickwork, see Fig. 5.3.

5.3 Ceramic Tiles and Ceramic Mosaic

Coloured glaze was part of the repertoire of decoration in the architecture of the ancient Middle East, in both Egypt and Mesopotamia, but the time Islam arrived on the scene, these techniques had been long forgotten. The earliest Islamic monuments made extensive use of mosaic as both floor and wall decoration, but these were very much in the still-flourishing tradition of late-Antiquity. It was not really until the 12th century that architectural ceramics began to be used extensively in an Islamic setting (by way of the transmission described above). The gradual development of suitable ceramic glazes, in both pottery and tile-work, represented a whole series of technological advances, and as such was as much a scientific/technological achievement as an artistic one. The enthusiasm for the intense colours produced by these techniques meant that they were eventually transmitted right across the Islamic world, from the

Fig. 5.4 Example of cut tiles. This example is from the Alhambra. The coloured scroll work is similar to the edging used in Roman mosaics (Patterns in Islamic Art web site 2017, SPA 1203)

Atlantic to Central Asia. Ceramic tiles are found in most Muslim countries, using an extensive range of techniques that include high moulded relief, polychrome, lustreware and sgraffito. In addition, sophisticated techniques were developed that used pieces of cut tiles, bonded together with plaster, to form elaborate, multi-coloured mosaic panels. The latter method was particularly favoured in the lands that came under Timurid influence (Iran and Central Asia), and in the Islamic far-West (Spain and Morocco). The broad area that fell under Iranian influence had a long, independent tradition of ceramic architectural revetment, known as *kashi*, which reached its first great achievement in the 14th century and was sustained through the Timurid and Safavid periods. Each element of the traditional Islamic decorative canon, geometric and vegetal arabesque forms, together with calligraphy, is used but with great local variations of style. The use of cut-tile mosaic seems to have come into favour in the Islamic West (the Maghreb) with the appearance of the Berber dynasties in the 12th-13th centuries. Known as *zellij*, it is almost exclusively dedicated to geometrical arrangements. Moorish Spain and Morocco had long followed a linked cultural existence, somewhat independent of the rest of the Islamic world, developing their own characteristic architectural and decorative forms. The particular specialty here, of intense geometric patterning is a tradition that has lasted right up to the modern period.

For examples of tilework, see Figs. 5.4 and 10.9

Fig. 5.5 Stucco/Plasterwork from the Alhambra (Patio de los Arrayanes)

5.4 Stucco/Plasterwork

Plaster was a well-established building material prior to the Islamic conquests both in Iran, where it had been used to cover rough, rubble walls for centuries, and to a lesser extent in the Classical, Mediterranean world. This was a readily available material in the Middle East, and was used in Islamic architecture from the very earliest periods (in Syria and Iraq), spreading fairly rapidly to the rest of the Muslim world. Its earlier forms followed late-Classical and Sassanian models, but Islamic tastes were soon asserted, inclining towards a flattening of the decorated surface, the emphasis on symmetry, its division into distinct, evenly laid-out panels, and the use of abstract rather than naturalistic motifs. Plaster, a singularly useful material that leant itself to being moulded and carved in a variety of ways, became a staple of Islamic architecture. Perhaps because of its plasticity as a medium it was less frequently used for purely geometric designs, and was more often used in vegetal-arabesque arrangements. It could also, of course, be painted or gilded. In essence, plaster (on its own or in conjunction with ceramics) was the perfect medium for transforming surfaces—which was always the primary concern in Islamic architecture. As indicated earlier, the aesthetic intention in Islamic art generally is bound up with ideas of the dissolution of matter, of transcendence.

For an example of stucco/plasterwork, see Fig. 5.5.

5.5 Woodwork

Since wood is a comparatively scarce material in many parts of the Islamic world it perhaps not surprising that it enjoyed a higher status as a material than elsewhere and, at its best, displays the very highest levels of workmanship. Traditionally, it was used for doors and window shutters, which are frequently inlaid, but the finest work

Fig. 5.6 Wooden minbar from Mosque of al-Mu'ayyad, Cairo. The pattern has a 12-pointed star and an exact pentagon (Tiling Search Web Site 2017, data162/P072)

is generally found on minbars, the key piece of furniture in the mosque from which Friday sermons are preached. Many highly sophisticated techniques were developed to create intricate decoration; in the finer examples ebony and other precious woods are used as inlays, together with ivory and mother-of-pearl. The carving in these objects often has a concentrated, almost lapidary detail. In the Islamic world the skills of carpentry were traditionally associated with geometry. Some surviving examples of 12th century woodwork indicate that the genre of complex, interlacing geometrical designs in the *girih* mode were relatively common by this time, and may have been expressed in this architectural medium before any other (after their probable invention as Qur'ān illumination in the 10th century; see below).

It appears that early woodwork used a complex procedure of interlocking pieces which does not require metalwork or glue. This is called *ta'sheeq* (Arabic), *ta'siq* (Persian) or *kündekari* (Turkish). For instance, the 15th-century Victoria and Albert minbar (Victoria and Albert Museum collections website 2017, 48775) seems to have no metalwork. The reconstruction of the minbar of Saladin (Singer 2008, pp. 172–173) used this technique with photographs of the assembly. Although the individual panels are held together this way, the V&A minbar can nevertheless be taken apart, a curator stated: rather like a 15th-century version of Ikea flat pack furniture.

For examples of Mamluk woodwork, see Figs. 5.6 and 14.8.

5.6 Book Illumination

As indicated above, the earliest appearances of the girih mode are found in late 10th century Qur'ān manuscripts that are believed to have originated in Baghdad. It is thought likely that this style spread out to other media from this source. Although the precise mechanisms by which this style was transmitted are obscure, obviously it would have been facilitated by the widespread availability of paper soon after this time. The three components of the Islamic decorative canon (calligraphy, geometrical and vegetal arabesque arrangements) which were established in this medium continued to be a feature of the Islamic arts of the book, with every period making its own contribution to the genre. As the most elevated of all objects the Qur'ān naturally always received the most respectful and inspired treatment, but there are also

Fig. 5.7 Anonymous Baghdad Qur'ān. This is the right half of a doublepage opening to a volume of the Qur'ān. Its patron is unknown, but we know that the calligrapher was Ahmad ibn al-Suhrawardi, a famous student of the master scribe Yaqut al-Musta'simi, and that the illuminator was Muhammad ibn Aibak, who collaborated with him on several outstanding Qur'ans. From the surviving parts of this Qur'an, it appears that the illuminator created a different geometric scheme to preface each volume. The Metropolitan Museum of Art (50.12)

many examples of geometrical patterning and arabesque in secular book illustration. Clearly, the art of decoration was taken seriously in Islamic culture; patterns had a currency. Their appearance in the portable arts (that is the art of the book, in craft-albums and working drawings), meant that existing examples could be reworked and improved upon, and those who specialised in decorative patterns could inspire new directions in this way. There are countless examples of intricate pattern in each of the major traditions of miniature painting (Turkish, Persian and Mughal Indian), in fact, they are positively teeming with them. The exploration and playing with decorative ideas that is evident in so many of these works is probably indicative of a general interest in the subject in studios of various kinds. Sadly, apart from these examples and the few surviving folios and working drawings, the originators of the majority of these brilliant patterns and arabesques are likely to remain obscure (Fig. 5.7).

For an examples of an illuminated Qur'ān, see Figs. 1.1, 11.3a and 13.7a.

5.7 Court Patronage and Cottage Industry

There is a distinction between the artistic and luxury goods that were commissioned and made in centralised, royal studios and workshops, and those that were essentially the products of small-scale cottage industry. This division, however, between centralised and independent manufacture in traditional Islamic cultures, was often a rather more complex business than this simple apportioning might suggest. It was usually the case that rulers and their court custodians were the chief arbiters of taste and fashion in all things, but in most periods there were thriving manufacturing centres of such goods as textiles, pottery, metalwork and glassware, with their own design traditions. Many fine articles were designed in royal studios and manufactured in royal workshops exclusively for use by the Court, and occasionally these products were made available to other wealthy clients. In other cases designs were created in an official studio and the work then commissioned from independent artisans, many of whom worked in districts close to royal palaces. Religious institutions were of course another major source of commissions. These differences of origination, together with differences in local tastes and traditions, account for the greater variation of design in the arts of 'everyday objects', by comparison with the consistencies of the formal decorative modes in architecture. Symmetry, as always in Islamic design, was still key, but the traditional decorative canon (of geometrical patterns, arabesque arrangements and calligraphy), while an ever-present and general influence on design, exerted less of a hold beyond architecture. The design studios retained by Islamic ruling houses tended to exercise stylistic control and consistency across a whole range of media, including architectural decoration. In many case this stylistic continuity is fairly evident—the designs on Nasrid silks, for instance, have clear similarities with contemporary stuccowork. Textiles, partly because of the important role they played in courtly and ceremonial life, were frequently a focus for court designers, and textile workshops themselves were actively encouraged by the ruling establishment in many periods. There are abundant examples of this sort of stylistic conformity across the arts in the Ottoman, Safavid and Mughal dynasties, and it is safe to assume that this was the case in earlier periods. There is evidence from various Islamic periods of complex designs being prepared in design studios and passed on to skilled artisans across a range of luxury items, including (at different times), metalwork, carpets, glasswork and ivory-carving. In fact a great many artistic products from the Islamic world, in all media, are characterised by highly organised arrangements with complex schemes of patterning that clearly indicate a prior design stage. The essential role of paper in this process has already been touched on (in Chap. 4). Although very few examples survive, it seems extremely likely that the use of working drawings (and the inevitable appearance of pattern books and design folios), led to the creation of a design 'currency' among skilled artisans which enabled the continuous refinement of patterns, and their exchanges between media.

Chapter 6
Countries and Regions

6.1 Egypt

Egypt was brought into the Islamic sphere in the first wave of conquests, in 639. With the Tulunid assumption of power (868–905) it became an independent centre. Then, under the Fatimid dynasty (909–1171) it assumed its enduring role as the cultural focus of western Islam. The architecture of the Mamluk period (13th-16th) saw a continuity of earlier Egyptian traditions, but it also incorporated aspects of Iraqi and Syrian styles. Most of the buildings of these periods are of stone; in fact the general feeling conveyed by medieval Cairene architecture is that of a sombre stolidity (Fig. 6.1).

This is reflected in the various forms of architectural decoration. Of particular interest are the massive carved masonry domes, the rich polychrome marble inlays and painted ceilings. The somewhat severe beauty of this architecture is frequently relieved by the exquisitely carved and inlaid woodwork of doors and minbars.

6.2 India

India was partially conquered by the Arabs as early as the 8th century. The influence of Islam was maintained over successive centuries until, by the end of the 16th century, the greater part of the Indian sub-continent was under the rule of the Islamic Mughal emperors. In its earlier periods Indo-Islamic art and architecture borrowed heavily from Hindu forms, but over time a distinctive Indian-Islamic cultural style was forged (Fig. 6.2).

There is little that survives of early architecture, but the Qutab mosque and the extraordinary Qutab Minar in Delhi give some idea of these past glories. The majority of the geometric patterns from Indian sources presented here, however, are from the Mughal period. Despite the fact that Indo-Islamic architecture represents, to a greater

B. Wichmann and D. Wade, *Islamic Design: A Mathematical Approach*, Mathematics and the Built Environment 2,
https://doi.org/10.1007/978-3-319-69977-6_6

Fig. 6.1 This marble inlay panel is from the al-Maridani Mosque, Cairo, dating from 1340. Other examples of the same pattern can be found in Iran, Samarkand and India (Patterns in Islamic Art web site 2017, EGY 1609)

Fig. 6.2 Carved masonry and stone relief, Agra Fort, Agra. See also (Fig. 14.3a). This example is dated 1573, with earlier examples in Iran and Cairo. This pattern is unusual in having a 6-way intersection (Patterns in Islamic Art web site 2017, IND 0332)

extent than in any other region, a fusion between the pre-existent tradition (Hindu) and Islamic style, all the classic forms of Islamic architectural decoration are all well represented here. The exquisite inlaid marble patterns of the Tomb of Itmad ud Daula make it an architectural gem par excellence. The polychrome panels of the tombs of Humayun and Akbar are also of great interest, as are the many red sandstone decorative panels that grace the beautifully preserved city of Fatehpur Sikri.

Fig. 6.3 Ceramic mosaic panel from Fatima's Haram mosque, Qom, Iran. Date uncertain. Other known examples of this pattern are found in Iran (Patterns in Islamic Art web site 2017, IRA 3032)

6.3 Iran

The Persian Empire succumbed to the newly Islamicised Arabs in their early expansion (mid-7th century), immediately after the conquest of the Byzantine Middle East. But this ancient culture, though subjugated, was never entirely assimilated, and in time the Persians came to influence and contribute more than any other race to the building of Islamic culture. The Mongol, Timurid and Seljuk dynasties all left important monuments in Iran, but the majority of the examples here date from the Muzzafarid (1314–1393) and Safavid (1501–1786) periods.

Glazed decoration was a feature of Persian architecture from the earliest times and is found in most periods, but it assumed an increasingly important role under Muslim patronage, transforming the buildings to which it was applied (Fig. 6.3). Muzzafarid and Safavid art revel in sumptuous decoration, their architectural decorations use brilliant colours and highly evolved decorative schemes to virtually 'dissolve' the buildings to which there are applied. Apart from their ceramic tiles and mosaics there are rich traditions here of decorative brickwork and carved stucco. The mosques of Isfahan and Yazd show particularly fine examples of architectural decoration.

6.4 Morocco

In the westernmost part of the Islamic world, Morocco has always retained a certain national separateness. Its distinctive artistic and architectural styles evolved more or less independently of the countries of the Middle East, but are strongly connected

Fig. 6.4 Ceramic tiles, mosaic and pottery, Tomb of Moulay Ishmael, Meknes. 18th century, see (Patterns in Islamic Art web site 2017, MOR 0808). No other instances of this pattern has been found

to those of Islamic Spain. Five major dynasties have ruled Morocco during the past twelve hundred years, each leaving their own architectural legacy in their own capital city. The principle surviving monuments are in or near the towns of Fez, Meknes and Marrakesh. In its typical forms, Moroccan architectural decoration uses complex geometric ceramic mosaics in combination with intricately carved stucco. Finely carved and painted woodwork (particularly doors) are another common feature (Fig. 6.4).

All three modes of the Islamic decorative canon (geometric, arabesque and calligraphy) are expressed in a characteristic, somewhat intense style. This is a decorative art that makes a powerful impression, but it is on a human scale. Some of the finest examples of this somewhat cerebral ornamentation can be found in the various medieval madrasahs.

6.5 Spain

Spain had an Islamic presence for some eight centuries, and for the three of these it was important enough to have its own Caliph, based in the Umayyad capital at Cordova, quite independent of Baghdad. After the fall of the Umayyad dynasty (in mid-11th century) a series of minor dynasties ruled, until the Nasrid dynasty re-established political and cultural unity. At their court in Granada the Nasrids

Fig. 6.5 From the North end of Sala del Mexuar. Note how intricate the design is, including the central khatem at 45° to the others (Patterns in Islamic Art web site 2017, SPA 1919)

created a magnificent culture that recaptured some of the earlier glories. But this last stronghold of Islam in Spain was itself finally overcome by the Christian monarchs Ferdinand and Isabella in 1492 Hispano-Islamic art and architecture, in both major periods, was marked by an extreme aesthetic refinement (Fig. 6.5).

Miraculously, enough has survived in the Nasrid palace of the Alhambra in Granada to give us some idea of this marvellous art. Here practically every surface is decorated. Each wall has a dado of intricate ceramic mosaic, displaying a dazzling range of geometric patterns; above this, every wall surface is covered with moulded and carved plaster, creating a seemingly endless variety of arabesque forms, all of the panels are interwoven with bands of calligraphy. Although there is a definite consistency in the decorative scheme the overall effect is far from repetitive; it is perhaps the most exquisite architecture in the world. The architectural decoration of the Alcàzar in Seville, actually commissioned by the Christian King Pedro I, still resonates with the taste and workmanship of Muslim artist/craftsmen (see Sect. 1.11).

6.6 Syria

Syria, with its deep Classical and Christian cultural heritage, was Islam's first encounter with an advanced, urban civilisation. In fact, the conquest of Byzantine Damascus (in 635) was a defining moment in Islamic history, and the first Muslim dynasty, the Umayyads, soon made Damascus their capital. The Great Mosque in Damascus, the earliest mosque to have survived more or less intact, is interesting for precisely these reasons; the mosaic decorations are in the Byzantine tradition, but reflect a new Islamic artistic sensibility, and the same leavening influence on late Classical forms is evident in other architectural features, such as panels and window tracery (Fig. 6.6).

Fig. 6.6 Panel in semi-precious stone inlay featuring a 14-pointed star from the Jaqmaqiye madrasah, Damascus. This unique design features in (Bourgoin 1879, Plate 165)

In 1401 Damascus was occupied by Tamerlane, suffering terrible destruction as a result, an event from which the city never really recovered. A century later Syria was conquered by the Turks, was absorbed into their empire and became an Ottoman province. There are, nevertheless, many interesting monuments in Syria, particularly those dating from the medieval period—the Sabrinye madrasah and the mosque of Taynabiye in Damascus; the Umayyad mosque, the al-Otrush mosque, the madrasah al-Faradis and the mashad al-Hussein in Aleppo, are all impressive and important from an art-historical perspective.

6.7 Central Asia

Long before the region had become part of the Islamic world, the river Oxus in Central Asia had always been the traditional boundary between civilisation and barbarism. For the first three centuries of Islamic rule the barbarian nomads of the steppes beyond were held in check, as they had been under the Persians, and the cities of Bukhara and Samarkand prospered. But by the end of the 10th century the infiltration of Turkish nomads, the ‘Turkish Irruption’, had began in earnest. This great movement of peoples, highly important for the future of Islam was, by turns, a highly destructive and an extremely creative force.

Fig. 6.7 Ceramic tiles, Samarkand, the Tuman Aka mausoleum, part of the Shakh-i-Zindeh complex, 1405 (Patterns in Islamic Art web site 2017, TRA 0214)

Very little architecture remains from the older Islamic civilisation of this region, but the Mausoleum of the Samanids (early-10th century), whose surfaces, both internal and external, are entirely patterned with decorative brickwork, is an architectural masterwork. In fact this highly original building set an architectural style that was to be enormously influential throughout the eastern Islamic world. Because of their geographical location the cities of this region were always in the front line of nomad incursions, and the two greatest of these, the Mongol and Timurid, had a devastating effect on the region. But there was a pattern of recovery; even these ferocious hordes were eventually civilised. In fact Timur made Samarkand his capital, and his descendants established a brilliant and influential culture. The majority of the surviving monuments in Central Asia are in the style established by the Timurids, or their successors, the Shaybanids. This architecture is characterised by its monumental proportions and richness of its decoration, with entire surfaces being covered with ceramic tiles or glazed brick. Among the most impressive monuments are those of the Shakh-i-Zindeh (Fig. 6.7) and Registan complexes in Samarkand.

6.8 Turkey

Anatolia was conquered and brought into the Islamic fold by the Seljuk Turks in 1077. As was the case with other dynasties of nomadic origins, they bought a new vigour and originality to the arts and architecture. They were the first in Islamic architecture to use ceramic decoration on a monumental scale and the style that they pioneered in the 12th and 13th centuries became a source of inspiration, particularly in the eastern Islamic world, for centuries to come. As well as using ceramics on a grand scale the Seljuks made an extensive use of fine stone-carving to decorate their monuments. In both mediums they took the art of patterning to new levels of subtlety and intricacy.

Seljuk power declined at the end of the 13th century, and their successors the Ottomans introduced entirely new architectural and decorative forms rejecting many of the traditional notions, but some aspects of this tradition were retained, notably in their use of complex geometrical wood-inlay patterns in doors and other woodwork (Fig. 6.8).

Fig. 6.8 Carved and inlaid woodwork, Eski Mosque, Edirne from (Patterns in Islamic Art web site 2017, TUR 0121) 1414. Constructed from standard decagonal tiles, see Chap. 12

The stunning use of ceramic decoration in the Seljuk period is exemplified by the surviving interiors of the Alaadin mosque and Karatay madrasah in their capital, Konya. The splendid tradition of carved relief can be seen in the extraordinary Ince Minare madrasah (also in Konya), and in the many caravanserais strung out across Anatolia.

Part II
Mathematical Analysis

By Brian Wichmann

Based on two square ceramic mosaic panels on the east wall of the Salon de Comares, Alhambra Palace, Granada, Spain, between the three window alcoves. This composition is noteworthy in that the centre motif is at a slightly smaller scale than the rest of the panel, otherwise it would not fit.

Chapter 7
Introduction

The script is spiritual geometry, although made perceptible by a physical instrument.

Euclid, cited by Abu Baker as-Suli (Moustafa and Sperl 2014, vol. 2, p. 287)

The quote above comes from the monumental work by Moustafa and Sperl on Arabic penmanship. Their result is a mathematical analysis of the Proportional Script—an elegant form of calligraphy. The analysis undertaken here is much simpler than the curves in calligraphy since we restrict our attention to geometric patterns using straight lines. The importance of geometry and proportion in art was clearly understood in the early years of the Islamic world.

The Book of Wisdom project ensured that the works of Euclid were available to the Islamic world Sect. 1.6 and Al-Khalili (2012). Even the geometry used in this book hardly extends beyond Euclid, so is certainly adequate for the artisans undertaking the construction of the patterns.

Gülru Necipoğlu gives some useful background to Islamic geometric design. The importance of mathematics (for instance, group theory) is noted (Necipoğlu 1995, p. 71), but the depth of the analysis is insufficient to construct many patterns.

The approach here is to analyse the design of a pattern and represent this in mathematical terms; then the representation can be input into software to draw that pattern with virtually unlimited precision. Thus the design is in a modern format which can even drive a machine-cutting device (or 3-D printer).

7.1 Geometry

Most Islamic geometric patterns can be thought of essentially as tilings of the plane, that is, as tessellations of convex or non-convex polygons of various shapes joined together at their edges with no gaps or overlaps. The vast majority are repeating

B. Wichmann and D. Wade, *Islamic Design: A Mathematical Approach*, Mathematics and the Built Environment 2,
https://doi.org/10.1007/978-3-319-69977-6_7

patterns, in which the same minimal groupings of polygons are repeated over and over in at least two directions, potentially filling the whole plane to infinity. As such, each Islamic pattern must conform to one or more of seventeen ways in which plane patterns are allowed to repeat (Conway et al. 2008; International Tables for Crystallography 2006). Each of these seventeen ways forms a *symmetry group* and each has its own unique internationally recognized symbol, indicating the manner in which various sets of *symmetry operations* combine to produce each symmetry group. The symmetry operations describe the manner in which the elements of a pattern are allowed to repeat throughout the plane, whether by shifting a certain *translation distance* in a given direction; by reversing or reflecting across a *mirror axis*; by combining these last two in a *glide-reflection axis*; or repetition by *rotation* a certain number of times round a centre. It so happens that only 2-, 3-, 4- or 6-fold rotation centres are permitted in two-dimensional repeating patterns. As we shall see, many Islamic star motifs centred on such centres contain stars with higher numbers of points, but their higher levels of rotational symmetry are only of limited, local application, and cannot extend to the whole plane. The local symmetry of such star motifs must be a whole multiple of one of the permitted values for the four different varieties of centres. The seventeen symmetry groups are described in Appendix A with examples.

The designers of Islamic patterns would not have known about the seventeen forms of repeating pattern (Conway et al. 2008), but these forms are useful to distinguish and analyse patterns.

7.1.1 Rosettes

The iconic characteristic of Islamic patterns is that of the *rosette*. We define this as a central star surrounded by identical polygons as shown below. The illustration shows *kites* (the four-sided polygons with an axis of symmetry), but those may be omitted by removing the two small edges. The 6-sided polygons are called *petals* and have a mirror line. The mirror line extends to the centre of the star. In the illustration, the four outer edges of the petals have the same edge length (a common convention), and the two outer edges form part of the regular polygon; neither of these two properties are required for a rosette.

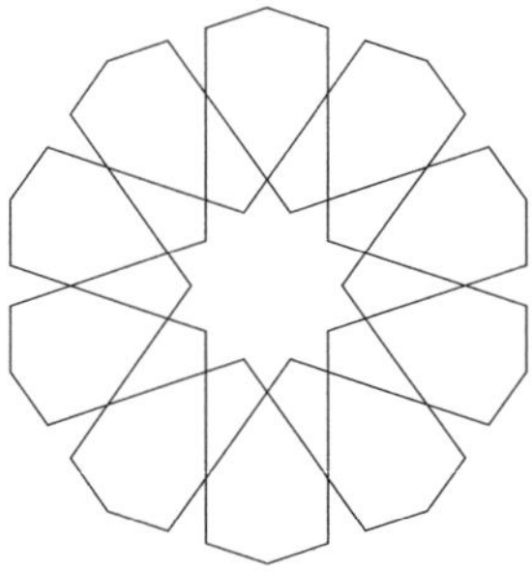

a

From a pair of minbar doors in the Metropolitan Museum of Art, 1325-1330 (Tiling Search Web Site 2017, data164/F294). Also a 14-th century Qur'ān in the Chester Beatty Library, Dublin

b

Widespread all over Islam. Earliest example is Qalawun, Cairo, 1284-5 (Tiling Search Web Site 2017, data10/P042). Occurs in different forms in the Alhambra

Fig. 7.1 Single rosette in a square formation

For the rosette above, the symmetry is *10• (*d10*)— having ten petals arranged round the central star, and the four outer edges of the petal have the same length. In many patterns, the rosettes form a major part; so we consider patterns in which rosettes are adjacent to fill the plane (albeit with gaps). The simplest arrangement is to have a single rosette arranged in a square formation. We show two such patterns in Fig. 7.1.

Figure 7.1a shows a graphic from an illustration of a Qur'ān, also shown as a photograph in Fig. 11.3a. The 12-pointed star has kites and is shown with up/down interlacing. The outer edges of the petal forms part of a dodecagon (as in the rosette above). Here, we are not considering the parts of the pattern outside the rosettes, usually referred to as the interstitial region. Again, the outer four edges of the petals have the same length, called a *standard petal*.

Figure 7.1b is a pattern which is widely used and appears in the Alhambra. Note that the central 8-pointed star does not form part of a rosette since the petals do not have a mirror line which goes through the centre of that star. This arrangement allows the introduction of a 4-pointed star. Considering the area centred on this 4-pointed

a

al-Fadawiyya dome, Cairo, 1479-1481, see (Tiling Search Web Site 2017, data8/PG268)

b

Minbar of the Sultan Shah mosque. Cairo, (Victoria and Albert Museum collections website 2017, 48775)

c

al-Ashraf Barsbay, Cairo, 1423 (Tiling Search Web Site 2017, data181/EGY1102). Carved Masonry & Stone Relief, see (Patterns in Islamic Art web site 2017, EGY 1102)

Fig. 7.2 Two rosettes in a square formation

star, one can see the pattern does indeed have a rosette, although the shape of the (blue) petals is different to the others we have seen so far.

Both these patterns have rotational symmetry of order 4 round the central star and also the same symmetry round another point. This other point is the centre of the octagon in Fig. 7.1a and the centre of the 4-pointed star in Fig. 7.1b. This implies the patterns as a whole have symmetry denoted by $*442$ (*p4m*).

A more complex arrangement is that of two different rosettes also in a square formation. Three examples are shown in Fig. 7.2. We consider these three examples together. All the rosettes contain kites, and the number of petals is either 8, 12, 16 or 20. None of the seven rosettes contain parallel sided petals. Note that Fig. 7.1b has a double ring of kites which is comparatively rare. There is an approximate matching of the widths of facing petals in Fig. 7.1c. The ratio of the sizes of the rosettes varies, while the visual impression is similar. All three have symmetry $*442$ (*p4m*), and hence the number of points are divisible by 4. The pattern Fig. 7.1c has been drawn with up/down interlacing, but the original does not seem to have that—the graphic can never be a copy of the original. In essence, the graphic tries to capture the concept of the design but not necessarily following every detail. In a similar way, the colour used in the computer graphics is not necessarily a copy of an original but designed to show the structure of the pattern.

a

Great Mosque, Divrigi, 1241 (Aslanapa 1971, Fig. 13); and Great Mosque, Thatta,1647 (Degeorge 2002, p267)

b

Alaaddin Mosque, Konya, 1116-1237 (Tiling Search Web Site 2017, data188/FIG5)

c

Al-Suhaymi House, Cairo (Tiling Search Web Site 2017, data12/P136). This design features in (Bourgoin 1879, Plate 136)

Fig. 7.3 Three arrangements with ∗632 (*p6m*) symmetry

Some further points about Fig. 7.2a. This pattern is taken from a Victorian publication and we have no modern photograph. This implies some uncertainty about the accuracy of the geometry. Notice also that all the cross-overs have four tiles meeting at a point which implies that the pattern could be coloured with just two colours. There is also a cross-over which is not straight—we consider this issue later.

We now consider three arrangements found with patterns in a hexagonal formation having ∗632 (*p6m*) symmetry. This implies a point with rotational symmetry of order 6 and another point with rotational symmetry of order 3.

Figure 7.3a is to be found in at least three places: the Great Mosque, Thatta, the Metropolitan Museum of Art (originally from Cairo), and the Great Mosque at Divrigi, Turkey. It is a simple honeycomb design. At the point at which three honeycombs meet, we have a tile which necessarily has 3• (*d3*) symmetry. This tile is convex, 6-sided, but not a regular hexagon.

Figure 7.3b is from Alaaddin mosque, Konya, in Turkey. Here the rosettes have no kites but the petals are bounded by a regular 9-sided polygon. The 6-pointed star is surrounded by unusual 10-sided polygons. This star is the centre of the 6-fold rotation.

Figure 7.3c is in the Al-Suhaymi House, Cairo. A very similar pattern it to be found in the minbar of mosque of Qijmas Al-Ishaqi in Cairo. The two stars have 9 and 18 points, both within rosettes. Note that the petals of the 18-pointed rosette narrows to meet the petals of the 9-pointed rosette.

a

Redouane Pavilion, Marrakesh (Tiling Search Web Site 2017, data 186/P119)

b

Ahmad Ibn Tulun, Cairo, 1296 (Tiling Search Web site 2017, data12/P143). This design features in (Bourgoin 1879, Plate 143)

c

Sevres porcelain, 1867 (Victoria and Albert Museum collections web Site 2017, 35011)

Fig. 7.4 Three octagonal arrangements with ∗442 (*p4m*) symmetry

In the arrangements in Fig. 7.4, we have eight smaller rosettes around the main star. Joining the centres of the 8-pointed stars gives anoctagon. The remaining space in the designs is a square which in all three cases is filled with an octagon surrounded by eight tiles. We will consider an example in the next Chapter very similar to the patterns Fig. 7.4a and Fig. 7.4c.

Note that rosettes with similar designs can nevertheless vary in detail. Filling the interstitial region to form a coherent pattern is part of the attraction of Islamic design which we consider in the rest of this book. Of about 1,300 Islamic patterns in the database at www.tilingsearch.org only roughly a quarter have true rosettes. We will return to the question of the statistics in Sect. 17.5.

Given a rosette, then the arrangement of the kites can change. In fact the rosettes considered above contain no kites (Fig. 7.3b), one ring of kites (Fig. 7.3c), or a double ring (Fig. 7.2b). A fuller treatment of rosettes is to be found in Lee and Soliman (2014).

Chapter 8
A Worked Example

In 1936 Maurits Escher visited the Alhambra with his wife Jetta and made a number of sketches (Schattschneider 1990, p. 17). One hundred years earlier, Owen Jones published his first work on the Alhambra (Jones and Goury 1837). Both subsequently produced a drawing of the same pattern from that building (Locher 1992, pp. 41, 53; Victoria and Albert Museum collections website 719669). Unlike Jones, we have the advantage of working from modern photographs, at least in this case (Fig. 8.1).

The pattern has the same symmetry as a square grid, that is $*442$(*p4m*) (see Appendix A, p 193); this is by far the most common symmetry for Islamic patterns. The arrangement of the two rosettes we have already seen with a very similar pattern in Fig. 7.4a. The same pattern appears in the Alhambra as part of a large ceramic mosaic dado occupying the whole of the north wall of the Sala del Mexuar each side of the entrance to the Mexuar Oratory, and the northernmost parts of the east and west walls. The continuity is interrupted on the west wall by the window opening onto the Machuca garden, and on the east wall by the exit to the Mexuar Patio. We have a record of only one instance of this pattern outside Spain and Morocco. Note also that the black petals with parallel sides mirrors the bounding square of the complete panel. In the Maghreb, square or rectangular panels are common, as are parallel-sided petals.

We are excluding from consideration here the border pattern (but see Fig. 11.4a for a complete pattern in this style). Borders are a common feature of Roman mosaics. Note that curves are not used, although it may give that impression.

The key to understanding Islamic patterns is careful study and analysis of the angles and relative lengths of the lines. Firstly, it is constructed from straight lines, the only other feature is the motif within the central star which we ignore.

The central star has 16 points, while the other stars have 8 points (the vertices being 90°). The black 6-sided polygons are petals which we have noted have parallel sides. Considering the angles of the kite marked X in Fig. 8.2 shows that the central star has a vertex angle of 45°. Since the sides of the black petals clearly extend to the petals of the 8-pointed star, it is clear the main pattern is highly constrained.

B. Wichmann and D. Wade, *Islamic Design: A Mathematical Approach*, Mathematics and the Built Environment 2,
https://doi.org/10.1007/978-3-319-69977-6_8

Fig. 8.1 Sala del Mexuar. The edge is ignored in the analysis of this pattern

There are four 6-sided figures at the four compass-point positions just outside the black petals. These four polygons appear to be the same shape as the petals of the 8-pointed star. The traditional way of drawing petals is for the four edges furthest from the star to be of the same length (as noted in the previous chapter). The 8-pointed star petals appear traditional while the 16-pointed petals are clearly not.

Hence the key point in our analysis is to assume that the compass point petals are *exactly* the same as the 8-pointed petals. We now need to compute all the lengths given that we know the angles as illustrated in Fig. 8.2. We can clearly take the length A as unity.

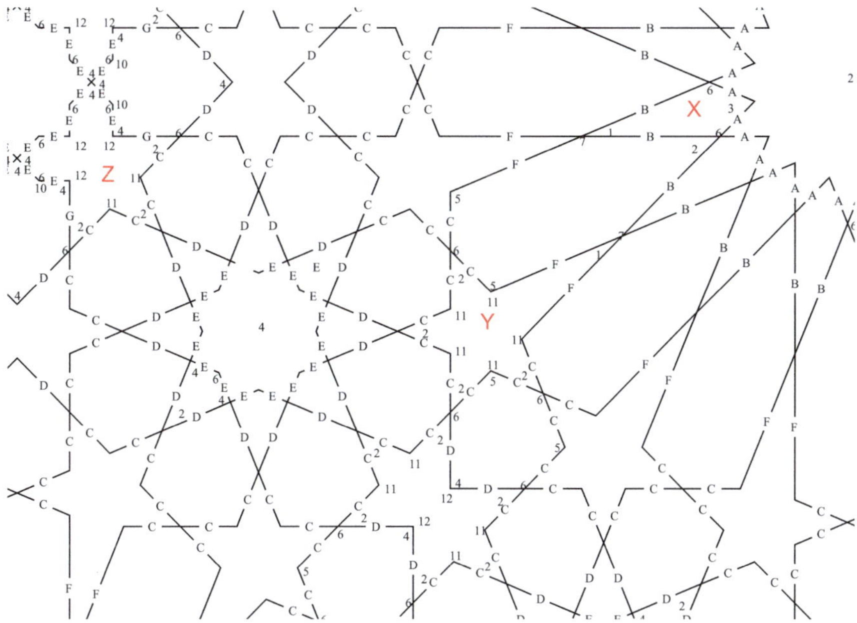

Fig. 8.2 Lines (A–G) and angles are multiples of 22.5°. (Multiples shown against some angles)

The main method of undertaking the computation is by repeated application of the sine rule: (angle)/(opposite side) is the same for all the angles of a triangle. For the first application we consider the triangle which is half the kite (marked X in Fig. 8.2). This gives:

$$A/\sin(11.25) = B/\sin(33.75)$$

giving the length B. The width of the main petals gives the length of C and E. The value of D follows by the same reasoning.

Formula
$A = 1.0$
$B = A * \sin(33.75)/\sin(11.25)$
$C = B * \sin(22.5)/\cos(22.5)$
$D = C * \cos(22.5)/\sin(45)$
$E = D * \sin(22.5)/\cos(22.5)$
$F = C * \sin(123.75)/\sin(11.25) - C - 2 * C\sin(22.5)$
$G = -2 * E * \sin(45) + 2 * C * \sin(45) - C\sin(22.5) + C * \cos(22.5)$

The formula for F is more complex, but there is little need to compute this. Consider the polygon marked with a red Y (in Fig. 8.2). All the angles are determined and the value of C is already known. Hence F is fixed. (Actually, software used to draw the patterns here computes F automatically.) Again, all the edge lengths are

Fig. 8.3 Final graphic, based on Fig. 8.1

known for the polygon marked with a red Z except G. Hence G is also determined (Fig. 8.3).

We have now captured this design to mathematical precision. Inserting this information into appropriate software, a 'perfect' graphic can be produced. Of course, we can be confident that the Alhambra craftsmen did not use the formulae above; however, given a size of the central star, the 8-pointed star petal could be produced.

Fig. 8.4 Occurs on arch jambs at entrances to east and west side chambers of the Sala de las Dos Hermanas. Colours of the graphic are approximately correct, but originals have alternating green and honey-brown rosettes. A similar mosaic remains on the floor between the arch jambs at the entrance to the Mirador de Lindaraja. This is now much worn away, and little of the original colour is left

By moving the ceramic tiles around, the size of the 16-pointed star petal would be clear (hence the value of F in our formulae). This technique can also be applied to produce the tile marked with the red Z.

Other variants of this design have been produced, including a ceiling in the Alhambra which has the same pattern in wood (Fig. 8.5). As is usual with such material, there is little colouring and the thick woodwork gives quite a different visual impression. Note that the variant shown in Fig. 8.4 was also drawn by Escher. There are two essential changes: the white interlacing is removed and an additional small square divides the 12-sided polygon in the original design, which we analysed above.

Fig. 8.5 Vestibule ceiling (Tiling Search Web Site 2017, data208/E41). Note that the central star actually has a vertex angle of 67.5° and extra small kites so that the rest of the design matches the other variants. The main star has two rings of kites, see also Fig. 7.2b

All these variants share the same property: 16 and 8-pointed stars with parallel sided petals. The width of the 16-pointed petal is the same as the width of the 8-pointed petal which then determines the size of the 8-pointed rosette. Two of the edges of the bounding octagon align with two of the petals from the 16-pointed rosette, thus effectively completing the design.

Chapter 9
An Octagonal Set

In this chapter, we are looking at not just a single tiling pattern, but a set of over 140 patterns. The set is based upon the khatem, the 8-pointed star so is called octagonal. Firstly, consider the three photos in Figs. 9.1, 9.2 and 9.3. These three patterns all come from Spain or Morocco; indeed, the whole set of patterns might well also be called of *Moroccan style*, see Paccard (1980).

Figure 9.1 is typical of ceramic work with coloured tiles alternating between dark and light colours. Here the light colour is essentially white, but the dark colour changes to increase the effect. Note here that the central stars (usually *khatems*) have a curved appearance to mimic the floral effect. Also, the pattern is not repeated in the obvious way at the edges.

For Fig. 9.1 we need to decide exactly what is the underlying pattern. Firstly, the obvious approach is to ignore the edge pieces—the large white tile and those further to the edge. Secondly, to take the khatems as having straight edges rather than the curves as in the central three (in a vertical formation). We are then left with these central khatems and others at the corners of a square with a 4-pointed star along each edge. This pattern has symmetry ∗442 (*p4m*). The graphic from this design has already been seen in Fig. 7.1b.

Figure 9.2 is typical of ceramic work with white up/down interlacing. This implies the use of dark colours for the main pattern — actually mainly black. Note that the larger black tiles have a yellow diamond added to increase the dramatic effect.

Figure 9.3 is typical of wooden work which is common on doors and ceilings. Here the tiles are painted with floral and geometric ornament; the up/down interlacing also apparent here but is much wider.

We wish to analyse these patterns in purely geometric terms. This means that some aspects of the patterns are ignored — such as the presence or not of the interlacing.

For Fig. 9.2, there is no need to ignore the edges since they can be seen as part of the pattern apart from the vertical edge ornament. We do however ignore the diamonds appearing in the large black tiles. This pattern also has symmetry ∗442 (*p4m*).

B. Wichmann and D. Wade, *Islamic Design: A Mathematical Approach*, Mathematics and the Built Environment 2,
https://doi.org/10.1007/978-3-319-69977-6_9

Fig. 9.1 Alhambra, Mirador de Lindaraja (Patterns in Islamic Art web site 2017, SPA 1009)

For Fig. 9.3 we have to decide the size of the repeated area. The khatems at the edge are truncated so we exclude them. Note that the symmetry is not the same as Figs. 9.1 and 9.2, since this pattern has no rotational symmetry of order 4, but is ∗2222 (*pmm*). We take the size of the pattern to be the repeated area bounded by four complete khatems (which appears twice in the photo).

In Fig. 9.4, we show these basic patterns produced by computer graphics so they can be compared.

These patterns share key properties:

1. The tiles are *edge-to-edge*. That is, neighbouring tiles share one edge which does not continue further for either tile.
2. The pattern can be coloured with just two colours.

Fig. 9.2 Bou Inaniya madersah/mosque, Meknes (Patterns in Islamic Art web site 2017, MOR 0913)

3. For each 4-way junction, the lines cross over. (There are only 2 and 4-way junctions)
4. For each tile, the internal angles are all a multiple of 45°. This is perhaps not so obvious, but can easily be checked.

There is another less obvious property concerning the lengths of the edges used. Most of these khatems are surrounded by kites. Take the length of the edges of the khatem as unity, the longer edge of the kites has a length of $1 + \sqrt{2}$. It is easy to see that the longer side of the large brown tiles in Fig. 9.4c has a length of $\sqrt{2}(1 + \sqrt{2}) = 2 + \sqrt{2}$. For the first pattern, see Fig. 9.4a, the edge length of the 4-pointed star is $\sqrt{2}$. These lengths cover all those used in the three patterns in Fig. 9.4.

Fig. 9.3 Alhambra, Patio de los Arrayanes (Patterns in Islamic Art web site 2017, SPA 0214)

We now produce our definition of the octagonal set. The set of patterns those satisfying the four conditions above and whose edge lengths are $p + \sqrt{2}q$ for integers p and q. However, if large negative values for p were allowed, any edge length could be obtained to a specified accuracy. Hence the only negative value allowed for p is -1. Over 140 patterns are known with the constraints applied here (see Figs. 9.5, 9.8 and 9.10).

We now show examples of these patterns drawn by computer, but following the original colours Fig. 9.6. The occurrence of these patterns in various parts of the Islamic world is shown in Table 9.1.

The most popular tiles amongst the octagonal patterns are shown in Fig. 9.10. This figure excludes the regular polygons and star-polygons (such as the khatem).

a b c

Figure 9.1. Early example is Qalawun, Cairo, Egypt, see (Tiling Search Web Site 2017, data10/P042). Alhambra date uncertain. Other examples in Iran, India and Morocco

In Fig. 9.2. All examples are from Morocco, including Bou Inania madrasah, Meknes, 18th century (Tiling Search Web Site 2017, data153/P164)

Figure 9.3. Alhambra, Patio de los Arrayanes, 1354-62 (Tiling Search Web Site 2017, data159/P077)

Fig. 9.4 Three patterns compared

a b c

Royal Palace Throne Room, Fez, unknown date (Tiling Search Web Site 2017, data152/C122B)

Royal Palace Throne Room, Meknes, 1672-80 (Tiling Search Web Site 2017, data152/C122C)

Fez, location and date uncertain, see (Patterns in Islamic Art web site 2017, MOR 0411)

Fig. 9.5 Patterns from Morocco

a

Attarine Madrasah, Fez, dated 1325. See (Tiling search Web Site 2017, data162/P058) Anexamplein Syria. Thisdesign features in (Bourgoin 1879, Plate 58)

b

Bou Inania Madrasah, Fez, dated 1351-6, see (Patterns in Islamic Art web site 2017, MOR 0203)

c

Alhambra, Sala de las Dos Hermanas, 1354-62 (Tiling Search Web Site 2017, data174/NC)

Fig. 9.6 Traditional Moroccan style

Table 9.1 Distribution of octagonal set patterns (see Chap. 6)

Region	Number	Percentage
Egypt	4	1.8
India	8	6.9
Iran	21	6.4
Morocco	54	21.1
Spain	38	18.3
Syria	2	3.4
Central Asia	9	6.8
Turkey	5	3.0

Table 9.2 Common edge lengths of tiles in octagonal set patterns

Edge	Formulae	Length	Example
A	1	1.0	khatem
B	$1+\sqrt{2}$	2.41421356237	Tile 1, Fig. 9.10
D	$\sqrt{2}$	1.41421356237	Tile 5, Fig. 9.10
E	2	2.0	Tile 17, Fig. 9.10
S	$2+\sqrt{2}$	3.41421356237	Tile 4, Fig. 9.10

a

Damghan, Iran. Minaret dated 1026-9. Other examples from Afghanistan, India and Turkey. This design features in (Bourgoin 1879, Plate 67)

b

Minaret of Jam, Afghanistan, 1190 (Tiling Search Web Site 2017, data15/A222). Other examples from Iraq, India, Iran and Syria

c

From Akbar's Tomb, Sikandra, 1613 (Tiling Search Web Site 2017, data172/IND0703)

Fig. 9.7 Indian patterns

a

No known source. This design features in (Bourgoin 1879, Plate 46)

b

Royal Palace, Marrakesh and Royal Palace, Ifrane. Dates not known (Tiling Search Web Site 2017, data152/C126B)

c

Alhambra, Patio de los Leones, 1370-6 (Tiling Search Web Site 2017, data190/ALH1)

Fig. 9.8 Two sizes of khatem

a

Moulay Ishmael Mosque, Fez, 1437. Note the 11-sided polygon (Tiling Search Web Site 2017, data152/C268A)

b

The David Collection, Spanish tile panel, 14th century (Tiling Search Web Site 2017, data184/DS4)

c

Location unknown (Tiling Search Web Site 2017, data157/P18A). Note black 8-sided polygon

Fig. 9.9 Patterns with less frequently used tiles

Table 9.3 Common tiles with usage (from 143) and examples. Just 43 patterns use only these tiles

Tile Number	Usage	Example
1	82	Figure 9.4a
2	65	Figure 9.6a
3	58	Figure 9.6a
4	53	Figure 9.6a
5	42	Figure 9.6a
6	41	Figure 9.7b
7	37	Figure 9.6a
8	30	Figure 9.6b
9	24	Figure 9.7c
10	24	Figure 9.9a
11	23	Figure 9.6b
12	21	Figure 9.6b
13	21	Figure 9.6b
14	17	Figure 9.7b
15	13	Figure 9.9b
16	11	Figure 9.7b
17	10	Figure 9.9c

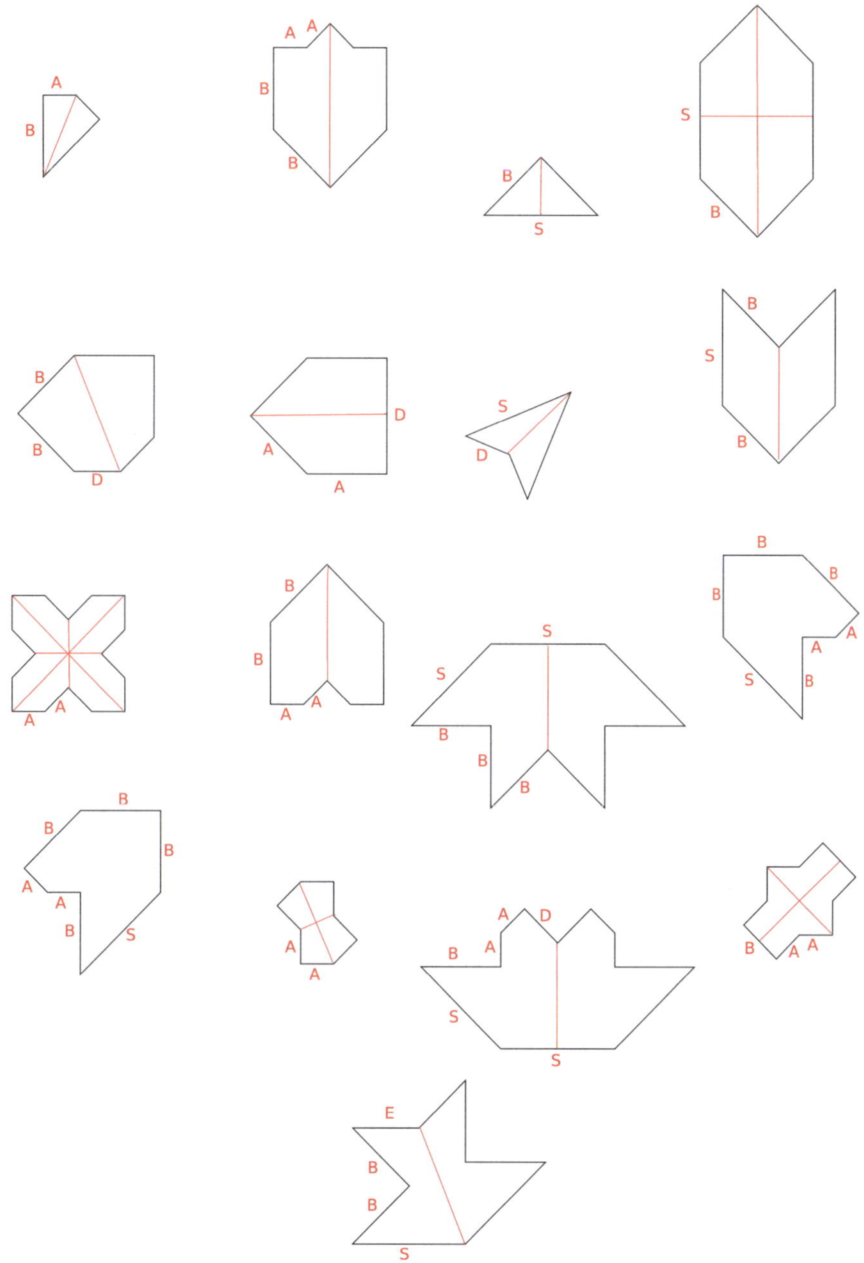

Fig. 9.10 The octagonal tiles used most frequently (1–17, numbered left to right, top to bottom) The red lines show the axes of symmetry. The red letters indicate the length of each edge. The total number of tiles used in the 143 patterns is 156

Fig. 9.11 Moulay Ishmael Mosque, Fez, Morocco (Patterns in Islamic Art web site 2017, MOR 0331)

The common edge lengths used are shown in Table 9.2. Of the total of 143 patterns in the octagonal set, 128 have edge lengths restricted to the values shown here.

The frequency and usage of the common tiling shapes in shown in Table 9.3. Again, the regular polygons are excluded, but the edge lengths show how the tiles can be fitted together (Figs. 9.8 and 9.10).

To conclude there are a large number of patterns in this set, mainly from Morocco and Spain. By construction, the individual tiles fit together to make patterns with ease. The colouring and possible interlacing adds further to their artistic appeal. Finally, we show a computer graphic of a complex pattern with interlacing Fig. 9.11.

Chapter 10
Octagonal Tiles with a Large Star

In this chapter, we consider a similar set of octagonal tiles to those of in the last chapter, but used to form a pattern with a large a central star. Three cases are analysed according to the number of points to the large star.

10.1 16-Pointed Stars

16-pointed stars are common in Islamic design, but appear in many different contexts. The simplest context is one in which the design is little more than the star with surrounding kites and petals, that is a rosette. We look at three examples of this in Fig. 10.1.

Consider the construction of Fig. 10.1a which is based upon a painted ceiling from Seville. As with Fig. 8.2, the vertex angle of the 16-pointed star is 45° so the petals have parallel sides. This implies that we can undertake the analysis and the computation of the lengths in the same way as in Chap. 8. To illustrate the specifics of this pattern, see Fig. 10.2. Starting with $A = 1.0$, we have $B = A * \sin(33.75)/\sin(11.25)$. The standard petals of the 8-pointed star are bounded by an octagon so the end petals of the large star are crossed at 45°. This gives $D = B * \sin(22.5)/\cos(22.5)$, from which E and G can be found. By considering the triangle outlined in red in Fig. 10.2, we can calculate C the edge length of the sides of the main petals.

The construction of Fig. 10.1b is actually straightforward assuming that the main petals are standard. Making the adjacent outer edges of the petals collinear is a frequent device, as in this pattern; therefore the main star is surrounded by a regular 16-sided polygon. This effectively determines the rest of the design. At the four points of the compass, the black darts join to the neighbouring rosette. This leaves a non-standard 8-pointed rosette with alternating petals; thus an octagon surrounds the rosette.

Figure 10.1c has an unusual property. The two long edges of the tile directly under the main star (marked E in Fig. 10.3) do not continue in a straight line (CE) to the next

B. Wichmann and D. Wade, *Islamic Design: A Mathematical Approach*, Mathematics and the Built Environment 2,
https://doi.org/10.1007/978-3-319-69977-6_10

a

Real Alcazar, Seville, also Sakkala, Tetouan, Morocco (Tiling Search Web Site 2017, data194/B129)

b

Ilyas mosque, Balat (Miletus), Turkey (Tiling Search Web Site 2017, data162/B57A)

c

al-Mu'ayyad, Cairo, 1420 and al-Ashraf Barsbay, 1423 (Tiling Search Web Site 2017, data160/ALW)

Fig. 10.1 16-pointed star with simple additions

tile. This is called an *interlace discontinuity*, and could be regarded as a defect in the design of a geometric pattern in Islamic style. To see how this has arisen, see the same diagram. The main petals are standard so the parallel edge lengths are determined. The six-sided polygon with edges C and D are also determined. The polygon marked Y is also determined since the only unknown is the value of F which can be calculated from the geometry. This implies that the width of the petals of the 8-star is determined, and then shows that the lines marked E force an interlace discontinuity. If this pattern were to be produced without the interlace discontinuity, then the 'border' would be much narrower which would give an unbalanced appearance. Note that if the edges marked E were straight, then the 6-sided dart-shaped polygons would not touch each other.

From the above, we have a conflict between the properties of an interlace discontinuity versus having the same width for the two types of petals (and hence having standard petals). For a further analysis of producing rosettes, see (Lee and Soliman 2014).

10.1.1 Adding Octagonal Tiles

Here we want to look at the general question of having a 16-pointed star with a background consisting of the octagonal tiles noted in Chap. 9. Of course, we have already met an example with an exact fit in Fig. 8.3. In that figure, the 32 symmetric six-sided polygons are in a context which gives an exact pattern.

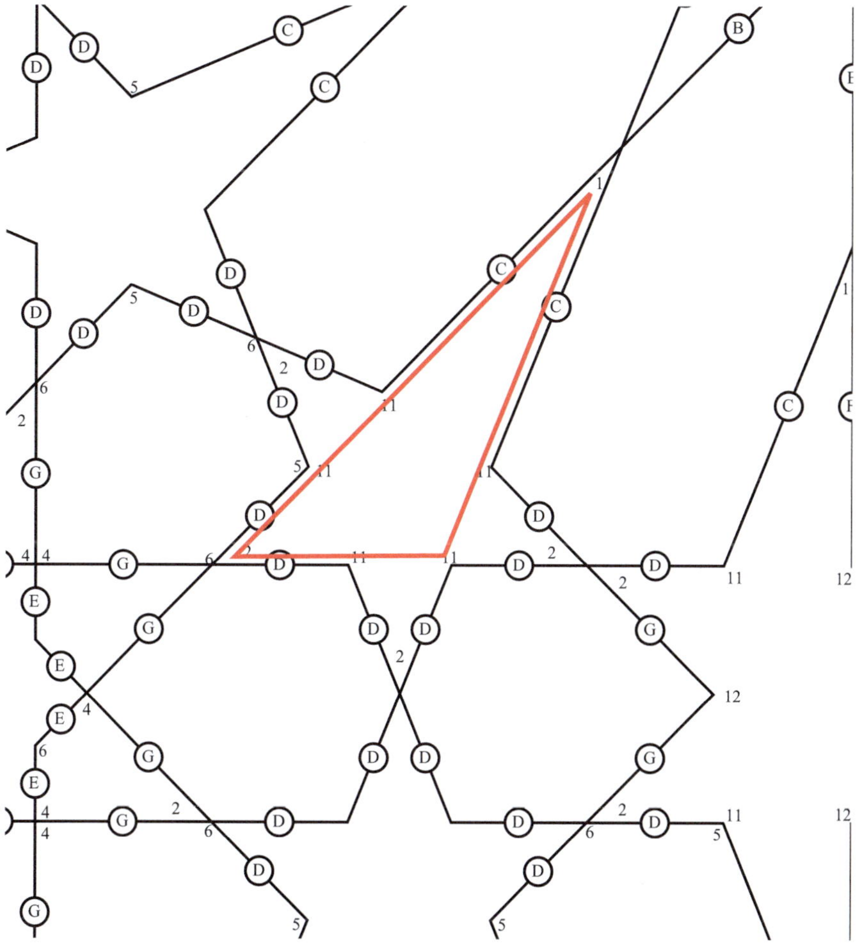

Fig. 10.2 Real Alcázar ceiling (part). Again, some angles are shown as multiples of 22.5°

Now consider the pattern in Fig. 10.4 in which the context is slightly different. Figure 10.5 shows there is a slight mismatch between the 16-pointed star and the octagonal tiles (whose relative size is fixed). One can see from Fig. 10.5 that this mismatch is quite small which implies that any adjustment needed is not visible from actual artefacts or photos. On the other hand, to get a mathematically precise pattern, we need to calculate an adjustment. (Indeed, the software used to produce the computer images in this book needs an exact calculation.) It is doubtful whether any of the original craftsmen have ever been aware that a mismatch exists; if it were brought to their attention, very likely they would regard it as irrelevant to their concerns.

The strategy is to compute a slightly different shape for the two six-sided black polygons either side of the black polygon directly under the main star (marked with red U and V in Fig. 10.6). Of course, these two polygons are mirror images of each

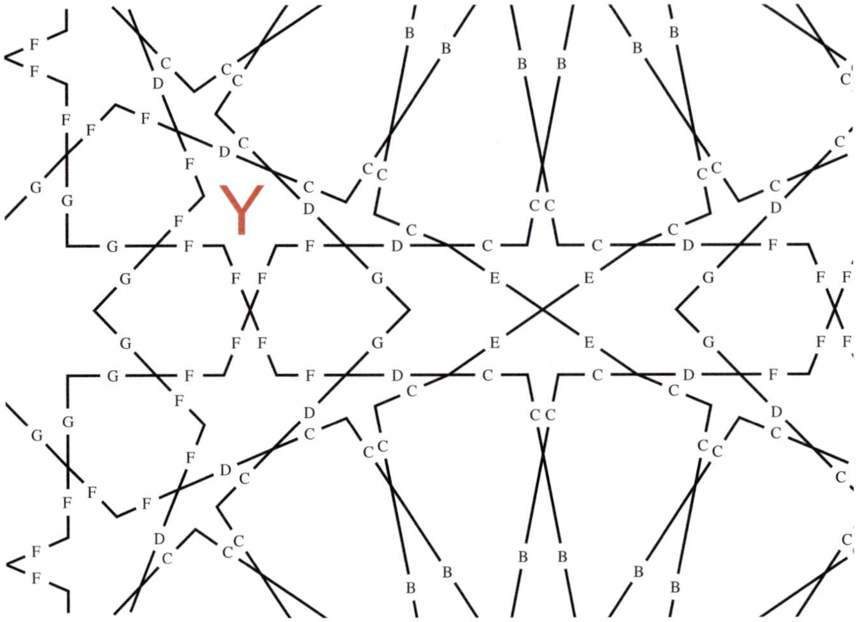

Fig. 10.3 Diagram of Sultan Barsbay Funerary Complex (part)

other. We must first compute the edge length of the parallel sides of the petals of the main star. This distance between the centre of the main star and the centre of the square directly under it is the same as that from that square to the khatem at the 4-fold symmetry point. This distance only involves octagonal tiles and therefore can be computed. In fact, the value is $(2+\sqrt{2})(\mathrm{B}+\mathrm{S})-\mathrm{B}$ using the values given in Table 9.2 which we denote as P. (Note that the edge length of the small square, not shown in Fig. 9.10 is $\sqrt{2}$, but see Fig. 10.5 and Table 9.2.)

Figure 10.6 shows the computation of the remaining edges, X and Y. Firstly consider the green triangle which extends to the centre of the star at the top. Hence the vertical edge is of length P. Using the sine formula we have:

$$(\mathrm{R}+\mathrm{B})/\sin(11.25)=\mathrm{P}/\sin(123.75)$$

which gives *R*. Now apply the sine formula to the red triangle:

$$\mathrm{R}/\sin(11.25)=(\mathrm{L}+\mathrm{R}+2*\mathrm{Q}*\sin(22.5))/\sin(123.75)$$

which gives L. Finally, apply the sine formula to the blue triangle:

Fig. 10.4 Moulay Idriss Zaouia, Morocco (Tiling Search Web Site 2017, data166/P155)

$$\begin{aligned}&(\mathrm{L}+\mathrm{R}+2*\mathrm{Q}\sin(22.5)+2*\mathrm{B})/\sin(112.5))\\&=(\mathrm{Y}+\mathrm{S})/\sin(22.5)\\&=(\mathrm{L}+\mathrm{R}+2*\mathrm{Q}*\sin(22.5)+\mathrm{X})/\sin(45))\end{aligned}$$

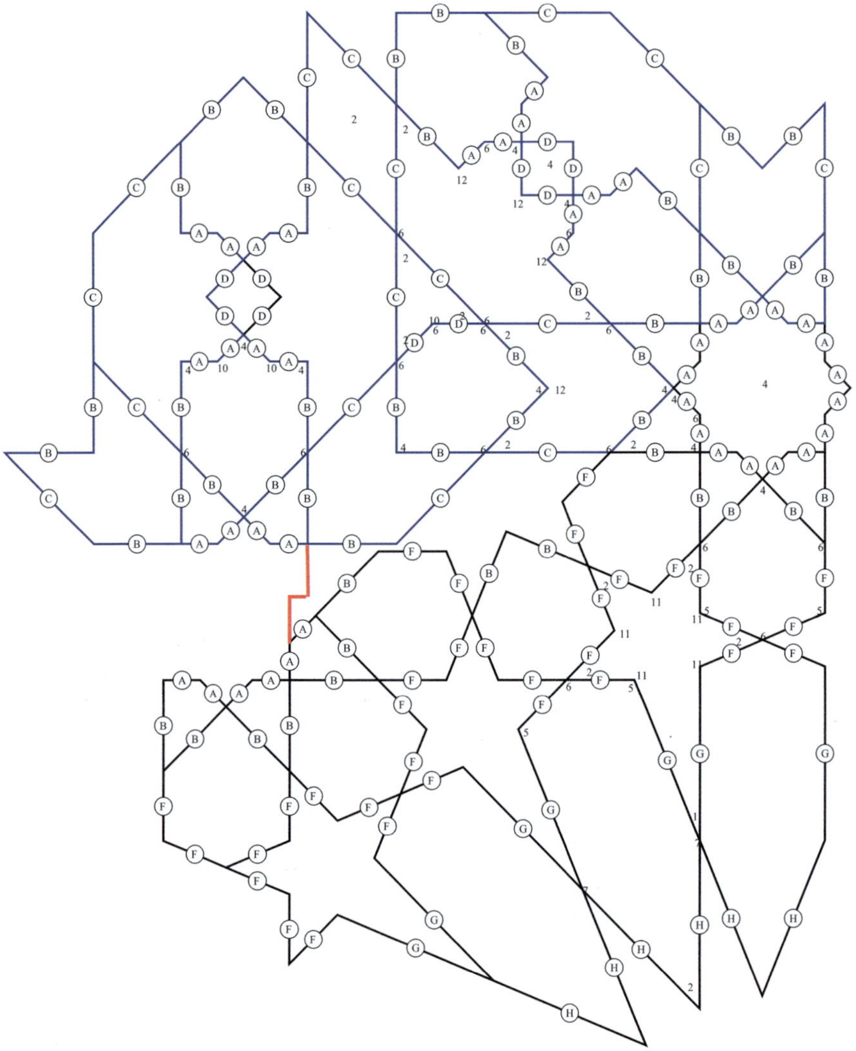

Fig. 10.5 Small mismatch between 16-pointed star and octogonal tiles; the blue tiles are octagonal tiles, while the other are derived from Fig. 8.3. The kink in the red line shows the mismatch

which gives both X and Y. Note that the key to such a mathematical analysis is the sine rule.

We conclude this section several patterns with a 16-pointed star having an octagonal background, Figs. 10.7 and 10.8.

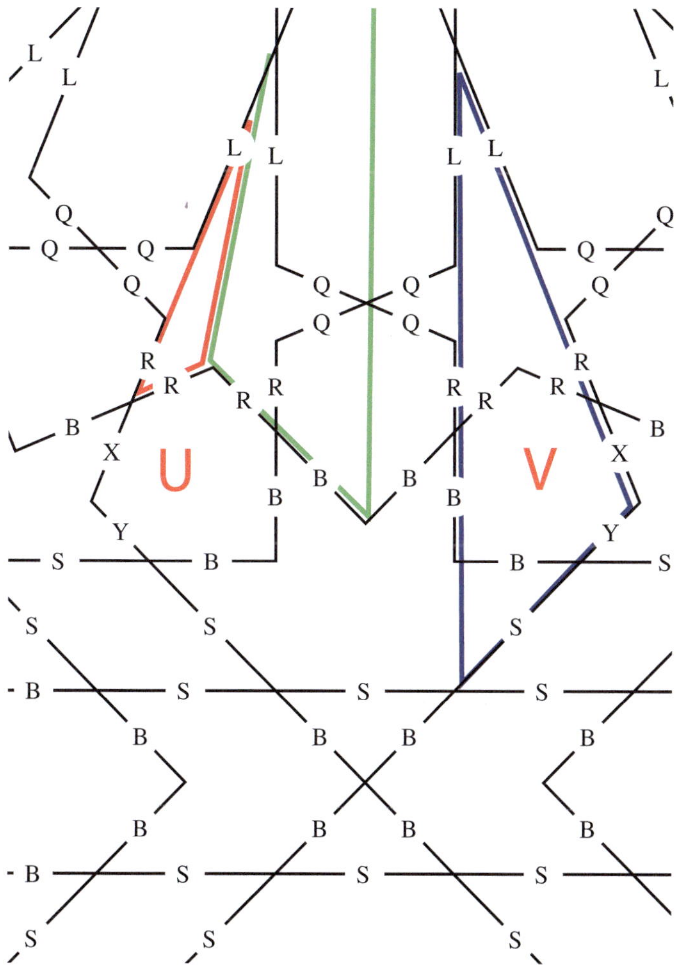

Fig. 10.6 Computation of edges

10.2 24-Pointed Stars

We have seen that with the 16-pointed star, the adjustments needed to allow the octagonal tiles to be used in the surround are minor and virtually invisible. The situation with the 24-pointed star is very different; this mismatch between the star and the octagonal tiles is much greater. This implies that the Moroccan craftsmen must have been aware of this mismatch and made numerous adjustments to overcome the problems. In our case with a mathematical description to be used for computer drawings, a simpler approach is needed.

a

Royal Palace, Marrakesh, colours added (Tiling Search Web Site 2017, data166/F14)

b

Attarine madrasah, Fez, 1325, original colour (Tiling Search Web Site 2017, data189/MOR0328)

c

Sefferine madrasah, Fez, 1271, original colour (Tiling Search Web Site 2017, data189/MOR0405)

Fig. 10.7 Moroccan examples

Here we take a typical *zellij* construction with a 24-pointed star in Fig. 10.9. This photograph shows significant damage presumably due to age, but it also shows substantial differences from a mathematical version which is now considered. To undertake our construction, we need to identify those tiles which can be octagonal and therefore of fixed size; then locate the tiles near the main star which need to be adjusted appropriately.

In Fig. 10.10 part of the star as shown—the tiles on the left have edge lengths of A, B and S using the lengths of octogonal tiles which were illustrated in Fig. 9.10. The X and Y mark polygons which are mirror images of each other; similarly with P and Q. This is a consequence of the pattern continuing above the red line by reflection along it. The vertical line under the end of the left hand edge of the red line marks a bounding octagon of the main star. The size of the main star is determined and therefore the lengths are marked with a 4 and 5 in Fig. 10.10. Now consider the right-angled triangle whose vertical left-hand edge is length $2 * S + B$. The horizontal bottom edge has length:

$$B + F + 2 * C * \sin(22.5) + l(3) + l(4) + l(5) = (2 * S + B) / \tan(22.5)$$

This therefore gives us $F + l(3)$. What is the value of F? It can be varied by a small amount and retain the general design. The value in other contexts would be to choose $F = C$, but that would give a value much larger than K which is already determined, since the polygon bounded by edges J, K and G is fixed.

We can see from this that the resulting pattern is far from ideal. For the X polygon, the values of the two sides (I and G) are very different. The two sides marked K are

Fig. 10.8 Large pattern from Marrakesh (Tiling Search Web Site 2017, data195/N13)

Fig. 10.9 24-pointed star from Fez (Patterns in Islamic Art web site 2017, MOR 0607)

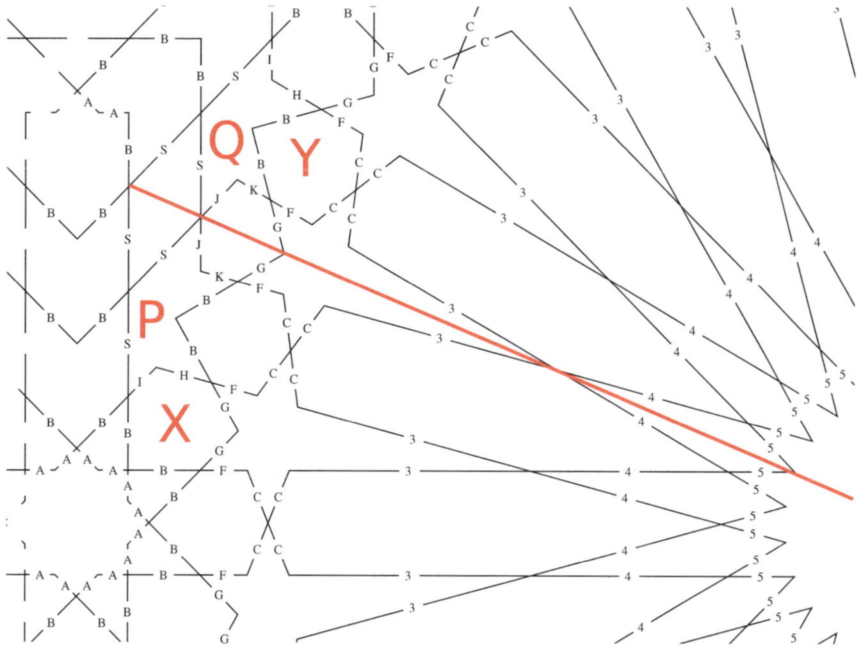

Fig. 10.10 Part of the 24-pointed star

not parallel, but visually, the reduced size of that polygon compared with X and Y is a flaw.

We can now compare the completed computer-drawn star in Fig. 10.11 with that of the photo in Fig. 10.9. It is clear that the central star was not produced by the original craftsmen with the vertex angle of 30° which is actually needed to get the parallel sided petals. This is hardly surprising for a cut ceramic tile. It appears that as a consequence, the petals are wider. Also the black 6-sided tiles do not vary in the way shown in the computer version. The conclusion is that the computer version deviates significantly but nevertheless has a beauty of its own.

From the active communities of craftsmen from Morocco (Paccard 1980, p. 382, vol. 1) and Iran, it appears that such designs are produced in a sand box. Individual cut tiles can be placed in the box and moved around. Further tiles can be cut and adjusted so that by a series of adjustments a completed pattern can be produced. At this point, the tiles can be fixed into their final positions.

We conclude this section by giving two instances of larger patterns with a 24-pointed star.

Figure 10.12 has a very large octagonal component, a 24-pointed star drawn in the manner we have described. Note the 8-pointed stars with a vertex angle of 45°. Also, there is an instance of tile 17 from Fig. 9.10 shown in green. There are also several octagonal tile shapes not shown in Fig. 9.10.

Fig. 10.11 Completed 24-pointed star (Tiling Search Web Site 2017, data190/MOR0607)

Figure 10.13 has both 16 and 24-point stars, both drawn in the manner described in this chapter. These stars are slightly different, extra ornamentation to the petals. This extra ornamentation still retains interlace continuity.

Fig. 10.12 Nejjarine Palace, Fez, 24-pointed star (Tiling Search Web Site 2017, data199/N6)

10.3 A 32-Pointed Star

In this section we look at a specific instance of the 32-pointed star which is surrounded by octagonal tiles. We have no detailed photograph of this tile, so we work from a completed graphic appearing in Fig. 10.14. This pattern was apparently designed by Maalem Moulay Hafid and is from the Royal Palace in Fez (Paccard 1980, p. 483, vol. 1).

As usual, a detailed study is needed. The brown tiles and those further outside the main star are octagonal tiles. Many of these tiles are shown in Fig. 9.10, but some additional tiles also appear. An interesting additional tile is the asymmetric 9-sided

Fig. 10.13 Madrasah Ben Youssef, Marrakesh, 16-pointed and 24-pointed stars (Patterns in Islamic Art web site 2017, MOR 0607)

Fig. 10.14 Royal Palace, Fez, 32-pointed star (Tiling Search Web Site 2017, data203/S32A)

brown tile directly under the ring of blue square tiles, appearing in both left and right-handed versions. An exception to the usual octagonal arrangement of the blue squares where the tile lines do not extend straight where to squares meet. Note also that the vertices of the yellow diamonds form a 6-way meeting of the tile edges. Such a 6-way arrangement appears only rarely when interlacing is used (for an example, see Fig. 14.3). These diamonds could be regarded as ornament to the petals of the main star.

The construction of this pattern can be followed Fig. 10.15. This shows the surround of the main star. The tiles marked X are from the octagonal set and therefore are of known size (and shape); so the main star's size is determined. From the tiles Y, the shape and edge lengths of the black tiles I and the white tiles R are determined.

Although we do not have a good photograph of this pattern, it is clear that the craftsmen would adjust tiles I and R to make the result appear more regular than is achieved by this mathematical version.

Other 32-pointed stars are quite different since the tiles I and R depend upon the configuration of the surrounding octagonal tiles. For another treatment of 32-pointed stars, see (Castéra 1999, p. 182).

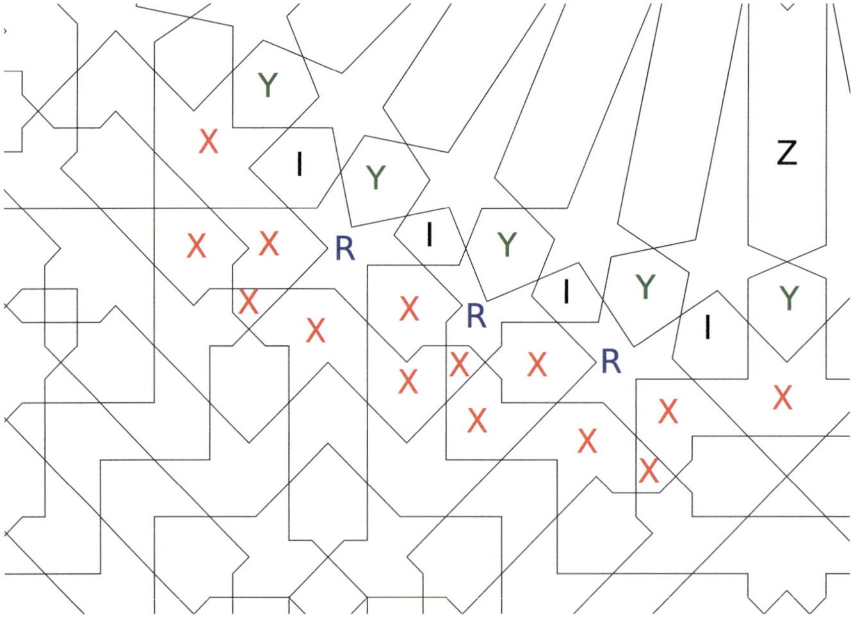

Fig. 10.15 Construction details, 32-pointed star

Chapter 11
Lines and Edges

The large set of patterns in Chap. 9 had the property edge-to-edge. This is the usual case, but there are exceptions (one example was shown in Fig. 5.3). Further examples arise when a tile is not even a topological disc, but has a 'hole' as shown by the red tile in Fig. 11.1c. We call this a *chelate*. The 'hole' is actually a small polygon (wholly within another polygon) which always seems to be a kite. Three examples are shown in Fig. 11.1. (Note that Fig. 11.1a has a similar arrangement as Fig. 14.5a.) All the examples here come from Cairo, and no example is known from the Maghreb. In spite of this irregularity, all the three figures could be coloured with just two colours.

In the rest of this chapter we consider patterns which are edge-to-edge, that is, any two adjacent polygons in the tiling always share a common edge between them. In this case, a certain number of polygons will meet at common points termed the *nodes* or *vertices* of the tiling. The number n of polygons meeting at a node creates an *n-way node*. Note that the number of edges meeting at that node is the same as the number of polygons. The number n may be the same throughout a tiling, or there may be different types of nodes. The sum of the angles meeting at any node must be 360°, although in general there is no restriction on the particular combination of angle sizes at each node.

One important category of patterns is of special relevance to the majority of Islamic geometric patterns, and this category comprises those that allow *interlacing*, as with Fig. 11.2a. With interlacing every node is 2-way or 4-way, and the distribution of angles at each node is such that opposite angles are equal, or what amounts to the same thing, adjacent angles sum to 180°. It also follows that opposite *edges* at each 4-way node are *collinear*, thus giving the impression of two straight lines intersecting at that node. The interlacing gives the impression of straight lines extending continuously through two or more adjacent nodes. This has given rise to the common means of representing these lines as straps or bands weaving alternately over and under one

B. Wichmann and D. Wade, *Islamic Design: A Mathematical Approach*, Mathematics and the Built Environment 2,
https://doi.org/10.1007/978-3-319-69977-6_11

a

Jamal al-Din al-Dhahabi, Cairo, 1637 (Tiling Search Web Site 2017, data201/F8785)

b

Aqbughawiyya, Cairo, 1340. Qibla wall, middle part of mihrab (Tiling Search Web Site 2017, data189/AJL1)

c

Amir Taz Palace, Cairo, 1352. Stone lintel (Patterns in Islamic Art web site 2017, EGY 0526)

Fig. 11.1 Examples of chelates

a

Itimad-ud-Daula's Tomb, Agra, date 1626 and Amber Palace, Jaipur, 1650-70 (Tiling Search Web Site 2017, data15/S27)

b

Panel from Tomb-marker, 14th century (Victoria and Albert Museum collections web site 2017, 109572)

c

Iranian, but no location or date (Tiling Search Web Site 2017, data197/IRAN11)

Fig. 11.2 Basic types of edge ornamentation

another, hence the name “interlacing patterns”. The ability to add such straps is a property of the pattern. (Actually, the present of chelates does allow interlacing as can be seen from Fig. 11.1b.)

Our mathematical approach is to think of patterns in terms of the basic structure of tiles and edges. In other words, we think of patterns in terms of their *base pattern*. We disregard colour and the manner in which lines are drawn. For instance, our computer drawings work from the base pattern so that for Fig. 9.1 the base pattern ignores the irregular part about the edge which is not present in Fig. 9.4a. Note that all the patterns in Chap. 9 have only 4-way nodes which are collinear. Since the computer can draw a line with any thickness, it is important to note that the base pattern is determined by the central line in the edge.

We now consider the various forms of edging. A very common form with patterns having 4-way nodes and collinear edges is that of interlacing, for example the three in Fig. 9.6.

Some artefacts are in wood which is often used in ceilings as in Fig. 8.5. There are many other splendid examples in the Alhambra. Note that due to the thickness of the woodwork, the up/down nature is much less obvious.

We consider three main forms of edge designs as follows:

Strict interlacing. This has borders which go alternately up and down. They are usually in white (but not always, as we shall see). The convention is to have interlace continuity. In the illustration in Fig. 11.2a, the up/down is clear, but with white-on-white this is sometimes less obvious.

Borders. Here we have simple borders, essentially wide edges, not just a line. In the case of wooden ceilings they are often very wide. See the illustration in Fig. 11.2b.

Banding. In this case we have a simple border between two tiles, but with a suitable polygon at the corners. See the illustration in Fig. 11.2c. The banding including the size of the squares can be varied. The banding could be reduced to zero in which case, the pattern would consist only of 6-pointed and 3-pointed star polygons.

11.1 Strict Interlacing

Figure 11.3 shows photographs of the three main types of interlacing: from a Qur’ān, a window lattice and a painted door. For the Qur’ān, the interlace has a yellow border which makes the up/down nature very clear. All the tiles here are elaborately ornamented (ignored in our computer drawings, see Fig. 7.1a). The window lattice is of high quality; many instances of this type are exposed to the elements and show significant damage. The painted door has both a border in brown and an up/down interlace. Sometimes this interlace work is very narrow, but this example is very clear.

a
Chester Beatty Library Qur'ān (here,1457, f.1a), from Egypt, 14th century, also Minbar Doors, Metropolitan Museum of Art, 1325-30

b
Amir Qijmas al-Ishaqi, Cairo, 1479-81 and al-Rifa'i mosque (here), 1869-1910. This design features in (Tiling Search Web Site 2017, data209/ALR1)

c
Royal Palace Museum, Fez, date unclear (Castéra 1999, p. 234)

Fig. 11.3 Photographs of examples of key types of interlacing

a
Alhambra, Fachada Solemne, 1354-62 (Patterns in Islamic Art web site 2017, SPA 0107)

b
Fritware from Fathbad, Uzbekistan, 1359 and 1360 (Victoria and Albert Museum collections web site 2017, 39498 and 102276)

c
Mir Chaqmaq mosque, Yazd, dated 1437 (Patterns in Islamic Art web site 2017, IRA 2004)

Fig. 11.4 Examples of multi-coloured interlacing

Fig. 11.5 Large multi-coloured interlace. Saadian Tombs, Marrakesh (Tiling Search Web Site 2017, data194/SA1). Note that the individual interlaces change colour

Examples of multicoloured interlacing appear in Fig. 11.4. Finally, a large example with colour changing Fig. 11.5.

a

al-Rifa'i Mosque, Cairo, 1869-1910, lattice design

b

al-Rifa'i Mosque, Cairo, kursi in prayer hall; and Amir Qijmas al-Ishaqi, Cairo (Creswell 1952/59, vol. 2, Plate 226). This design features in (Bourgoin 1879, Plate 83)

c

al-Mu'ayyad, Cairo (Patterns in Islamic Art web site 2017, EGY 1234). This design features in (Bourgoin 1879, Plate 132)

Fig. 11.6 Photographs of types of borders

a

al-Rifa'i mosque, Cairo, west wall of mausoleum of Sheik Ali Abu Shibak (Tiling Search Web Site 2017, data164/F6938)

b

Turbe of Kilic Arslan, Konya, 1156-92 (Degeorge 2002, p.179); Alaeddin Mosque, Konya, 1220, see (Tiling Search Web Site 2017, data166/179BM)

c

Ulugh Beg madrasah, Samarkand, 1420 (Hill 1964, Fig. 62) Ulugh Beg, Samarkand

Fig. 11.7 Examples of borders

11.2 Borders

Figure 11.6 shows photographs of three main types of borders: a lattice in wood (having effectively holes in the pattern instead of tiles), more elaborate woodwork (proud woodwork on a plain backing), and a door in metal. As is often the case, the petals of the main star in the metal door are proud of the main surface.

Figure 11.7 shows three computer images of a variety of borders: 11.7a is a panel in wood with a white background; the yellow colour has been added to emphasise the structure (the yellow tiles have added ornamentation in the original); 11.7b is a painted ceiling; both this pattern and 11.7c have pointed corners which encroach upon a neighbouring tile. The last pattern has thick grey borders, but the blue tiles have ornamentation in the original.

11.3 Banding

Banding is essentially a process of producing a new pattern from an existing one. Consider the example from Iran in Fig. 11.2c. In Fig. 11.8 we illustrate three patterns derived from the same design. Figure 11.8a is a pattern from Turkey with interlacing rather than banding. Figure 11.8b is the Iranian version with half width banding

a — Alaaddin Mosque, Konya, example with interlacing, 1155 (Aslanapa 1971, Fig. 87). This design features in (Bourgoin 1879, Plate 15 (top))

b — Example from Iran drawn with half width bands

c — Same example from Iran with double width bands

Fig. 11.8 Examples of banding

a

Alhambra, Patio de los Leones, 1370-6; also Toledo tile (Victoria and Albert Museum collections web site 2017, 160193)

b

Royal Palace, Maresh (Paccard 1980, p 390)

c

Another version with narrow banding

Fig. 11.9 Second example of banding

a

Huand Hatun mausoleum, 13th century (Hill 1964, Fig. 482); also in Iran

b

Gowhar Shad mosque, Mashhad, 1418 (Tiling Search Web Site 2017, data187/IRAN11)

c

Version with even narrower banding

Fig. 11.10 Third example of banding

a
Kok Gumbaz Mosque, Shahrisabz, Uzbekistan, 1437 (Tiling Search Web Site 2017, data12/P179-B)

b
Royal Palace, Fez (Tiling Search Web Site 2017, data199/FEZ1)

c
Version with very narrow banding

Fig. 11.11 Final example of banding

a
A building in Salisbury, and other places? (Tiling Search Web Site 2017, data2/P27B)

b
Cappella Palatina, Palazzo dei Normanni, 1132-40 (Hessemer 1990, p 112)

c
Roman mosaic, Gloucester. See (Neal 1981, p 86)

Fig. 11.12 Interlacing of defined width

compared to Fig. 11.2c. Figure 11.8c is a double width version. If the width were slightly larger, the 3-pointed star (with a vertex angle of 90°) would disappear.

We now look at a slightly more complex example of the same process in Fig. 11.9. We start with Fig. 11.9a, to apply the same process to obtain Fig. 11.9b. Of course, with the computer, we can again change the width of the banding, so in Fig. 11.9c we show a narrower version.

We increase the complexity. We start with Fig. 11.10a, to apply the same process to obtain Fig. 11.10b. With the computer, we can again change the width of the banding, so in Fig. 11.10c we show a narrower version. Note that the banded version is slightly different in that the bowtie shaped brown tile in (a) is now handled strictly according the banding process.

Finally for another example we start with Fig. 11.11a, to apply the same process to obtain Fig. 11.11b. Again, we can change the width of the banding, so in Fig. 11.11c we show a narrower version.

An example of a more complex form of banding is in Sect. 15.7.

11.4 Irregular Borders

We begin this section with three examples of interlacing that appear to be the same as before. The graphics are fine, but there is a problem; the width of the interlacing is defined by the geometry of the pattern and cannot be set to zero.

a — Crafts Center, Marrakesh (Castéra 1999, p. 99)

b — Toledo tile, see (Victoria and Albert Museum collections web site 2017, 160110)

c — Ghiyasiyeh Madrasah, Khargird, 1447 (Golombek and Wilbur 1988, Plate XI)

Fig. 11.13 Irregular banding

Figure 11.12a may well widely used, being simple and effective, but only two instances are known to us: a building in Salisbury and the Derby museum. Figure 11.12b is from the Royal Palace at Palermo which shows an Islamic style. Note that although the width of the interlacing can be reduced, making it zero changes the geometry almost beyond recognition. Figure 11.12c is actually Roman, but like (a) has a fixed width for the interlacing.

In Fig. 11.13a is unusual in having non-standard interlacing on one of the khatems. Although the wide of the white areas can be varied slightly, they cannot be reduced to zero. Figure 11.13b only has interlacing on part of the tiles. Figure 11.13c has polygons at the tile corners which vary in shape rather than being produced mathematically. As is frequently the case, the result shows artistic flare rather than strict adherence to the usual design principles.

Fig. 11.14 Path tracing, first example. Figure 9.5a, Royal Palace Throne Room, Fez. Note that pattern is in Moroccan style but is not octagonal in the sense of having tiles in Fig. 9.10. The red path is infinite but has only been traced in one direction. The other paths have 4-fold symmetry about the relevant two points of the pattern

11.5 Interlace Path Tracing

We conclude this chapter with three patterns having traditional up/down white interlacing, see Fig. 9.5. An interlace can be traced round the pattern until it either rejoins itself or repeats indefinitely. A problem soon arises: the path being traced goes outside the 'paper' being used! Computer software has been used to undertake the same

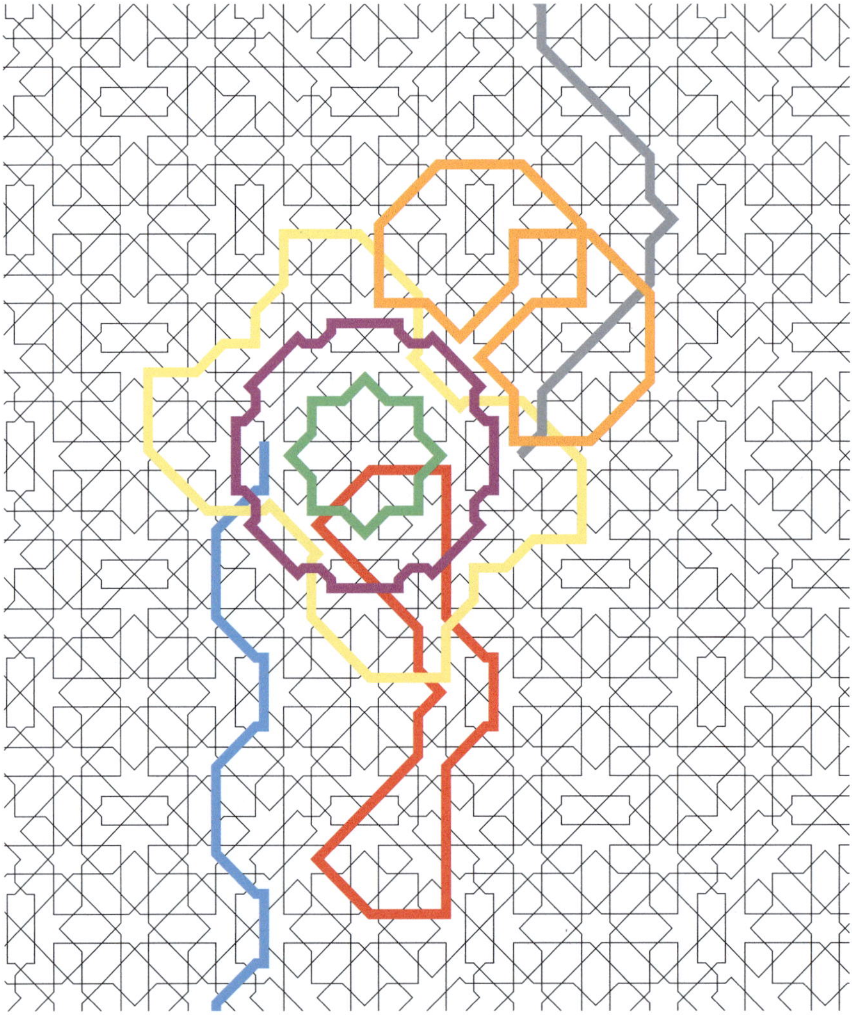

Fig. 11.15 Path tracing, second example. Figure 9.5b, Royal Palace Throne Room, Meknes. The grey and blue paths are both infinite. The red bow-tie path has an axis of symmetry corresponding to the pattern itself (similarly with the orange path). The other three paths have 4-fold symmetry about the relevant two points of the pattern

process, but with a large enough area to trace the paths. Grünbaum and Shephard (1992) noticed this problem. Their analysis of interlacing was manual rather than using software.

In Figs. 11.14, 11.15 and 11.16, the path tracing so produced is based upon three Moroccan tiling patterns already illustrated in Fig. 9.5.

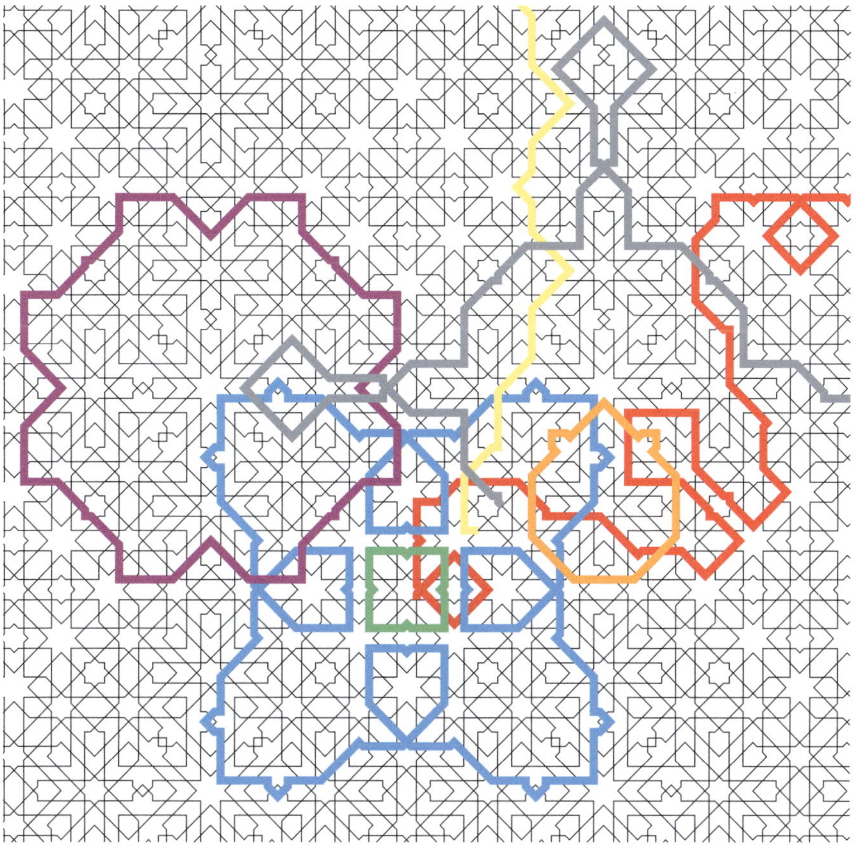

Fig. 11.16 Path tracing, third example. Figure 9.5c, Fez. There is one infinite path (yellow) and six finite paths. Note that the red and grey interlace paths are finite, but not shown in their entirety. The number of paths 1 (infinite) and 6 (finite) are invariants of this pattern but such numbers are too difficult to compute, in general, to be useful in distinguishing patterns

Chapter 12
Decagonal Patterns

Another large group of patterns similar to those introduced in Chap. 9 (octagonal patterns). A key property here is the presence of 10-pointed stars with a vertex angle of 72° or 108° as opposed to 8-pointed star with octogonal patterns. We saw that octogonal patterns came mainly from Spain and Morocco. Here, the patterns come mainly from Iran, Turkey or Egypt. As for the octagonal set, a few simple designs appear throughout the Islamic world.

12.1 A Simple Tile Set

This tile set consists of three irregular tiles, but the most important element is a 10-pointed star with a vertex angle of 108°. This star is always surrounded by kites. The other two irregular tiles have 8 and 10 sides. In addition, a regular pentagon appears in all the examples given in this section, five tiles in all.

The pattern in Fig. 12.1a is widespread but does not seem to appear in the Maghreb. Records show it present in Cairo, India and Central Asia (Tiling Search Web Site 2017, data12/P175). The colouring used is taken from the example at Samarkand, but adapted to show the geometric structure (since the original has extensive ornamentation within each tile) (Patterns in Islamic Art web site 2017, TRA 0211). Figure 12.1b is from Isfahan which has many excellent patterns, colouring is from the original. The Indian pattern in Fig. 12.1c is from Agra, produced using tiles cut from naturally coloured stones.

The patterns in Fig. 12.2 are all parts of a larger pattern; 12.2b appears in the corner of a large ceiling display, 12.2b has symmetry 5• (*c5*); 12.2b has reflective symmetry from the centre to the five corners and rotational symmetry of order five, ∗5• (*d5*), while 12.2c has symmetry ∗10• (*d10*).

B. Wichmann and D. Wade, *Islamic Design: A Mathematical Approach*, Mathematics and the Built Environment 2,
https://doi.org/10.1007/978-3-319-69977-6_12

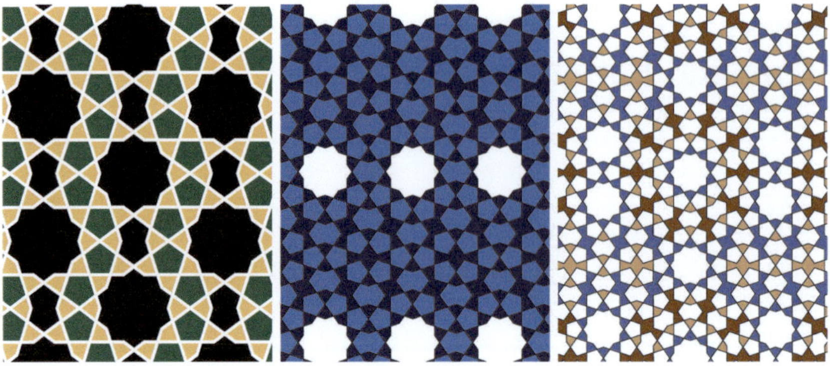

a

Friday mosque, Isfahan, 1088 (Seherr-Thoss, 1968). Widely distributed in Eastern Islam. This design features in (Bourgoin 1879, Plate 175)

b

Friday mosque, Isfahan, 1088 (Patterns in Islamic Art web site 2017, IRA 0532)

c

Itmad ud Daula, Agra, 1622 (Tiling Search Web Site 2017, data169/F6)

Fig. 12.1 A simple tile set

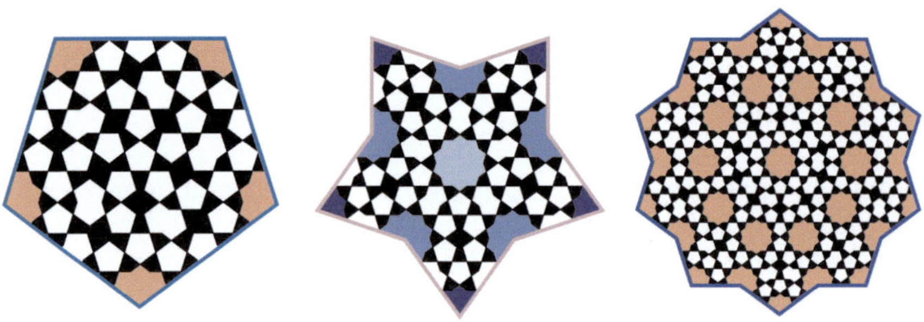

a

Darb-e Imam shrine, Isfahan, 1453 (Patterns in Islamic Art web site 2017, IRA 0907)

b

Ali Qoli Aqa mosque, Isfahan, 1122 (Tiling Search Web Site 2017, data187/IRAN9)

c

Darb-e Imam shrine, Isfahan, 1453 (Tiling Search Web Site 2017, data189/IRA0907B)

Fig. 12.2 Three roundels from Isfahan, Iran

12.2 Variable Diamond Patterns

One complication seen with decagonal patterns is that a few have diamonds which can vary in size; this also changes the size of the other tiles. This situation does not appear with the octagonal patterns in Chap. 9.

In Fig. 12.3, we show three examples of the same basic pattern. The main rosette was illustrated on p. 64. If one considers the central star including the diamonds, then it is clear that motif can be contracted by a small amount so that the pentagons and the 8-sided tiles (which look like two pentagons together) will be correspondingly larger. Figure 12.3a has the smallest diamonds and Fig. 12.3c the largest; the examples also show another variation with a change in the style of presentation. Note that Fig. 12.3c appeared in Fig. 11.11a as an example to which banding was applied.

In this case, the pattern has been produced with variable-sized diamonds. In other cases, the only artefacts known are with specific sized diamonds, in spite of the geometry permitting a variation.

To analyse the variation in geometric terms, consider Fig. 12.4. Taking the edge length of the 10-pointed star as fixed (A), then considering the angles of the kite, we have $B = A * \sin(54)/\sin(18)$. Since the petals are standard, we can compute the value of C. This gives $C = B * \sin(36)/\cos(18)$. Now considering the triangle in red, the value of $D + E = B * \sin(72)/\sin(36)$. We now can specify the variation by introducing the ratio $R = D/(D + E)$. Having determined both D and E, then F is determined which gives all the lengths in the pattern. Finally, we have in Table 12.1 the values of R, obtained by measurement of photos.

a

Huamt Hattun madrasah-mosque, Kayseri, 1178-1243 (Tiling Search Web Site 2017, data12/P179)

b

Muradiye Cami, Edirne, 1435-6

c

Kok Gumbaz Mosque, 1437 and Metropolitan Museum of Art (from Egypt)

Fig. 12.3 A common pattern with variable diamonds

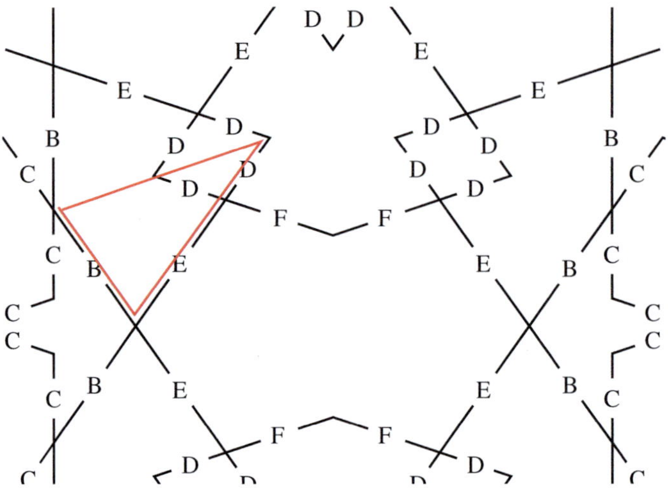

Fig. 12.4 Calculating the size of the diamonds

Table 12.1 Distribution of values of R for patterns in Fig. 12.3, (Tiling Search Web Site 2017, data12/P179)

R	Pattern
0.33	Serefeli Mosque, Turkey
0.45	Lid of Qur'ān box of Ottoman Sultan Selīm II
0.47	Havatan Turbesi - Tomb of Havatan, Istanbul
	Wakala al-Ghuri, Cairo in Egypt
	Hatuniye complex, Kayseri in Turkey
	The Metropolitan Museum of Art
	Edirne, Muradiye Cami (old door)
0.52	Kok Gumbaz Mosque
	& Dorut Tilyovat Complex, Shahrisabz

Figure 12.5a is essentially the same as Fig. 12.3a. The rosette of Fig. 12.3a is replaced by a 10-pointed star with a vertex of 108°, and has much smaller diamonds. But there is one other important change, extra diamonds are introduced next to the 10-pointed star, hence it is visually rather different. This pattern is from Mosque of Tan Sahid, Mashhad in Iran.

Figure 12.5b is also essentially the same as Fig. 12.3a. The change here is smaller, merely the removal of the kites round the 10-pointed star of Fig. 12.3a. Bernard O'Kane writes about this painted ceiling from Cairo. This is part of the sabil of a sabil kuttab, a building with a water dispensary (sabil) on the ground floor and a Qur'ān school (kuttab) on the upper. It was built by Abd al-Rahman Katkhuda, the Ottoman governor, in 1744.

a

Shah Mosque, Mashhad, 1629 (Tiling Search Web Site 2017, data197/IRAN6)

b

Abd al-Rahman Katkhuda, Cairo, 1744 (Broug 2013, p 177, Fig. 6.25)

c

From Cairo, (Victoria and Albert Museum collections website 2017, 1177879)

Fig. 12.5 More examples of variable diamonds

Figure 12.5c is a more complex pattern following along the same ideas. This is based on a drawing by James William Wild (date 1840) which is housed in the Victoria and Albert Museum. The drawing is remarkable since it was drawn years before the work of Bourgoin (1879) and yet shows an understanding of the design. The location of this pattern is unknown, but is surely Cairo, since he visited Cairo but no photograph is available.

12.3 Fixed-Size Diamond Patterns

There is a small set of decagonal patterns which have many diamonds and gives the impression that the size of the diamonds could vary.

For Fig. 12.6a, the values of A, B and C are the same as in the Fig. 12.5. Looking at the 9-sided polygon which has edges of length D and E only, an analysis shows that $R = 4 * \sin(18)/B$ and hence is fixed.

12.4 Adding a 5-Pointed Star

Here we consider decagonal patterns with a 5-pointed star with a vertex angle of 72°.

a — Rustem Pasha Mosque, Istanbul, 1561; and Suleymaniye Camii also in Istanbul

b — Suleymaniye Mosque, Istanbul, 1557

c — Darb-e Imam shrine, Isfahan (Patterns in Islamic Art web site 2017, IRA 0911)

Fig. 12.6 Examples of fixed-size of diamonds

a — Ali Qoli Aqa mosque, Isfahan, 1122 (Tiling Search Web Site 2017, data16/F114); also Royal mosque, Isfahan, 1611-29

b — Kasehgaran madrasah, Isfahan, 1105 (Tiling Search Web Site 2017, data187/IRAN4)

c — Shirdar madrasah, Samarkand, 1636 (Patterns in Islamic Art web site 2017, TRA 0409)

Fig. 12.7 Examples of patterns with a 5-pointed star

Figure 12.7a is from Ali Qoli Aqa mosque, Isfahan. The original is a motif on the wall with poor colour which has been enhanced in this graphic. Figure 12.7b uses the same tiles as (a) but is more elaborate. It is also from Iran—Kaseh-garan

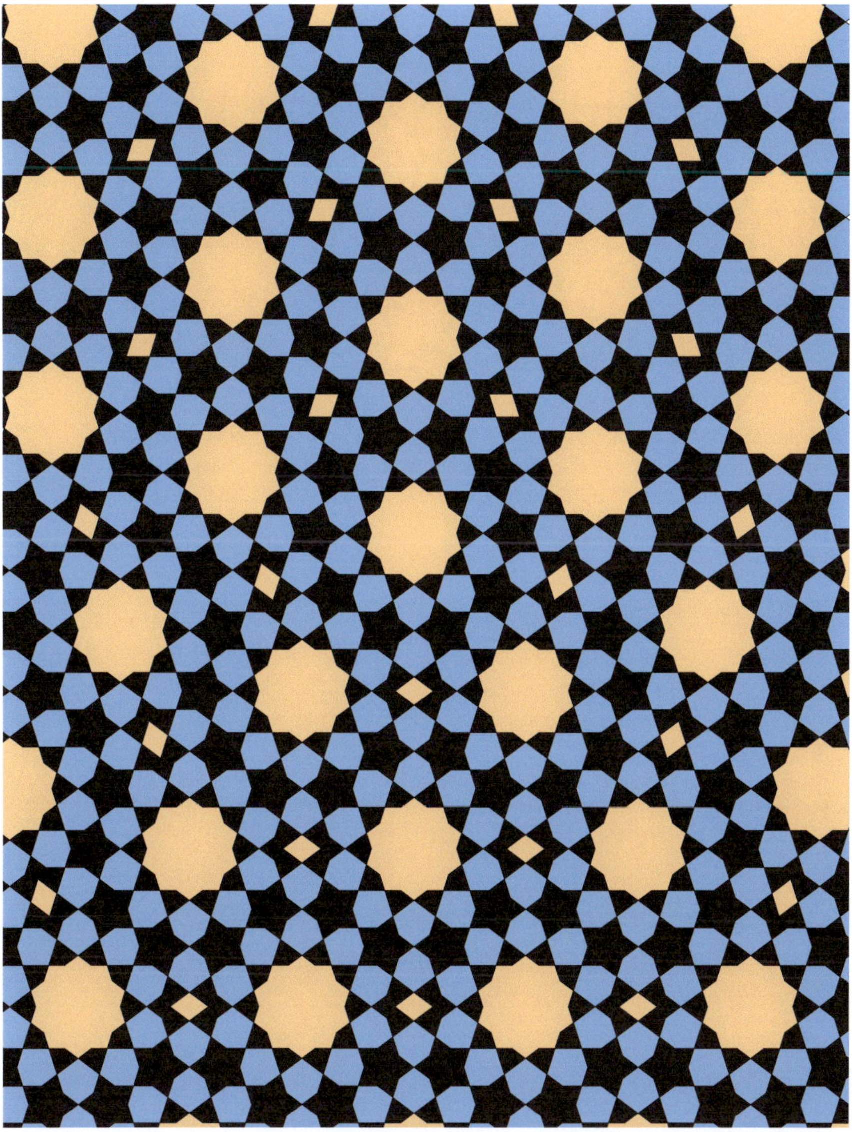

Fig. 12.8 A decagonal 5-pointed star pattern (Tiling Search Web Site 2017, data181/F111)

madrasah, Kashan. The original is a modest-sized roundel which has been extended here to a repeat pattern. Figure 12.7c is rather different since the polygons around the main star can be changed slightly—the shapes are not fixed in the same way as many decagonal pattern tiles. This pattern is from Samarkand.

Figure 12.8 is a roundel which has been extended using the symmetry $*5\bullet$ (*d5*). This shows a similar logic to that used by Kepler to extend a pattern. If one joins

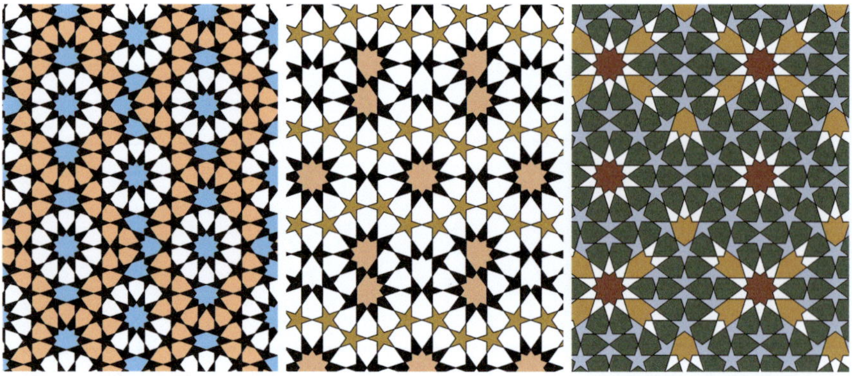

a
Madrasah-yi-Madar-i-e-Shah, Isfahan, 1714 (Patterns in Islamic Art web site 2017, IRA 1133)

b
Itmad ud Daula, 1626 (Tiling Search Web Site 2017, data185/P48)

c
Shabestan of the Friday mosque, Isfahan (Shafai 1977, p 69)

Fig. 12.9 Examples of patterns with a 5-pointed star

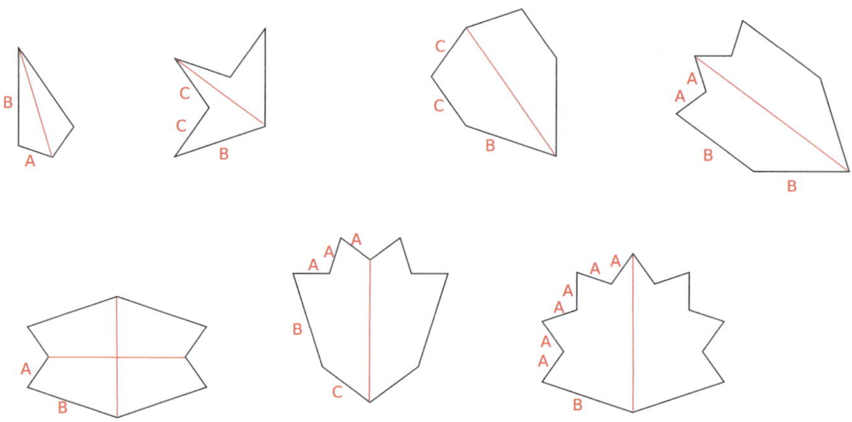

Fig. 12.10 A decagonal tile set (1–7, left to right, top to bottom). The red lines show the axes of symmetry. The red letters indicates the length of each edge

the centres of the 10-pointed stars, then the diamonds so formed with a central small diamond forms five wedges of diamonds from the central star (Fig. 12.9).

12.5 Another Simple Tile Set

This set is based upon the 10-pointed star with a vertex angle of 72°.

a

Abdulla and Modari Khan madrasahs, Bukhara (Tiling Search Web Site 2017, data197/DEC1)

b

Ghaznavid arch, Qala-i-e-Bust, Afghanistan, 1149 (Pope 1965, Photo 155); One of the most common, widespread and representative of all Islamic star patterns

c

Minbar from Friday mosque, Isfahan, 1310 (Tiling Search Web Site 2017, data198/BR5)

Fig. 12.11 Further examples of patterns with a 5-pointed star

a

The new courtyard, Mashhad (Maher al-Naqsh 1983, Vol 3, Fig. 103)

b

Topkapı museum, Istanbul (Tiling Search Web Site 2017, data198/TOP4)

c

An example of an old masterpiece

Fig. 12.12 Three erratic patterns

Fig. 12.13 A decagonal pattern with 5-pointed stars. This remarkable pattern uses 20-pointed stars as well as the conventional 5-pointed and 10-pointed stars. It is unusual to have three regular stars. From Bu 'Inanya madrasah, Fez 1355 (Tiling Search Web Site 2017, data195/FEZ1)

The tile set consist of seven irregular tiles (Fig. 12.10). Most patterns contain the 10-pointed star with a vertex angle of 72° and a 5-pointed star with a vertex angle of 36°. These tiles provide many possibilities for producing patterns.

Fig. 12.14 A pattern from Fez. This elaborate pattern uses mainly conventional decagonal tiles, but the use of the colour adds significantly to the effect. From Bu 'Inanya madrasah, Fez, 1355 (Tiling Search Web Site 2017, data167/P71)

The two patterns in Fig. 12.11a and Fig. 12.11c shows the variety and complexity that can be produced with this tile set. Figure 12.11b is one of the most common, widespread and representative of all Islamic star patterns 11th century onwards. A few examples occur in the Alhambra, such as a blind window grille on the otherwise blank south wall of the Mexuar Oratory; the wooden undersides of two support beams at the northern end of the Sala del Mexuar; and three window grilles in the Sala de los Ajimeces, above the entrance into the Sala de las Dos Hermanas. In the Generalife (Granada) the pattern occurs as carved stucco on the walls of the mirador in the north pavilion. Common throughout Moorish Spain.

Twenty examples of patterns using this tile set are known from Islamic sites. Of course, an infinite set of patterns can be produced.

12.6 Some Erratic Patterns

Figure 12.12a is a version of Fig. 12.1a in which each cross-over is replaced by four diamonds. No other example of this form of ornamentation is known to us. This is from the new courtyard in Mashhad according to the caption. Figure 12.12b is from the Topkapı Museum in Istanbul. Several examples are known of this type of design with the same complexity shown here. Figure 12.12c has two unusual features: firstly, the unusual tile shapes (the ones in green) and secondly that it is not edge-to-edge since the edge of the white pentagons joins to more than one tile. The caption from the Persian book reads ‘An example of an old masterpiece’ (Shafai 1977, p. 21).

Finally, we conclude this chapter with two examples from Morocco, a country not otherwise represented here (Figs. 12.13, 12.14).

Chapter 13
What Is Correct?

The previous chapter has introduced patterns which are relatively straightforward so their representation in mathematical terms does not present any difficulties. Here we consider examples in which 'Islamic style' comes into play.

13.1 An Example from the Alhambra

Our first example is a pattern which was drawn by Owen Jones as in Fig. 13.1 (Jones 1856, Plate XLII). The shape of the ceiling of the Sala de la Barca is essentially cylindrical, with quarter spheres capping each end. Consequently the basically planar pattern has to be greatly modified where it covers the spherical surfaces at each end. The colours on the painted wooden ceiling are black, white, silver, brown and orange.

Apart from the finished art work, Jones drew a sketch which may well have been produced on site at the Alhambra (Victoria and Albert Museum collections website 2017, 683705). (There is a slight problem here as this sketch is described as a wall decoration rather than a ceiling.)

Now contrast this with a modern photo and the graphic produced from this in Fig. 13.2. Careful inspection shows that the Alhambra pattern has an interlace discontinuity unlike the drawings by Owen Jones. When Jones visited the Alhambra sometime during 1832–1834, it would not have been possible for him to have used a camera. Hence it is not surprising that he made this mistake. (The problem is the line at 45° to the vertical from the 8-pointed star.)

In Fig. 10.2 we have already seen an interlace discontinuity which was forced by other constraints in the pattern. Does the drawing by Jones show that such constraints are not needed? Not really, since the adherence of Islamic style requires the rosette around a star to be as regular as possible. In the correct version, the rosette is bounded

B. Wichmann and D. Wade, *Islamic Design: A Mathematical Approach*, Mathematics and the Built Environment 2,
https://doi.org/10.1007/978-3-319-69977-6_13

Fig. 13.1 Ceiling of Sala de la Barca by Owen Jones (Patterns in Islamic Art web site 2017, JON 015)

a b

Fig. 13.2 Modern photo and the correct graphic of the ceiling of Sala de la Barca

by a regular 12-sided polygon. Hence, when objectives conflict as in this pattern, a choice must be made reflecting Islamic style. In the version by Owen Jones, the 6-sided polygon at the outer part of the rosette comes in three forms, while for the correct version they are all the same.

Verdict: Owen Jones missed the interface discontinuity which is easy to overlook.

13.2 Bourgoin Plate 32

We base our analysis on the drawing by Bourgoin (Bourgoin 1879, Plate 32). There is a pattern on a door in Aleppo which is similar but not very clear and seems to have some differences, so we will not include it in our analysis.

The essential difficulty is that of the heptagon and 5-pointed star. As the 5-pointed star has an internal angle of 120°, and the corresponding angle from the heptagon is 128.57°, we have a small discrepancy of 8.57°. Since both cannot be regular, so we have three options: Fig. 13.3a regular heptagon, Fig. 13.3b regular 5-pointed star or Fig. 13.3c neither but visually both looking regular.

For this plate, Bourgoin gives the details of his construction. He uses Fig. 13.3c, which we follow, is to make the vertices of the heptagon and the 5-pointed star circle inscribed. (Making nearly regular stars or nearly regular polygons circle-inscribed is a common technique, see Fig. 13.5.)

It has recently been brought to our attention by Goossen Karssenberg that the Friday mosque at Isfahan has a pattern with regular heptagons, as per Fig. 13.3a. The geometry of this pattern is not fully determined, and hence the actual match to the

a — Regular heptagon

b — Regular 5-pointed star

c — Visually both looking regular (as drawn by Bourgoin)

Fig. 13.3 The three variants of Bourgoin Plate 32

Fig. 13.4 Photo of the Friday mosque at Isfahan. The arch is in the Northern cupola with seven other panels

graphic here is poor as can be seen from Fig. 13.4. As far as we know, Bourgoin would not have seen this pattern.

Verdict: This pattern is essentially indeterminate. It seems best to follow Bourgoin and make both the heptagon and the 5-pointed star circle-inscribed as in Fig. 13.3c. The visual difference is small.

13.3 Bourgoin Plate 75

The collection of drawings produced by Bourgoin is probably the most widely referenced set (Bourgoin 1879). However, there is a significant problem with the collection: no sources are given, so that it is (in general) almost impossible to check the drawings.

For Plate 75, our guess is that the source is the mosque of Abd al-Gani al-Fakhri in Cairo. We know Bourgoin visited Cairo and many other patterns have that city as their source. Prior to locating the source, the drawing of Bourgoin was reproduced and so the photo can be compared Fig. 13.5.

a

Modern photograph

b

Computer graphic following Bourgoin, tiles with the same shape having the same colour

Fig. 13.5 Bourgoin, Plate 75

Looking carefully, one can see significant differences:

1. The black hexagon is regular in the drawing
2. The blue shape has a wider waist
3. The green petals are parallel-sided rather than slightly divergent.

A very close inspection shows that many of the straight lines are not quite straight which implies that care needs to be taken to produce a mathematical version which could be regarded as 'correct'. The width of the woodwork shows that slight interlace discontinuities can be accommodated.

In theory there is no single, inevitable construction for a pattern such as this, since the heptagon cannot be regular and other constraints such as interlace continuity cannot necessarily be satisfied. Hence that pattern must be produced in a way that is visually satisfactory, but lacks a mathematical basis.

The construction starts with deciding the vertex angle of the 12-pointed star. This must be a bit less that 60°, and a value of 52.5° matches Fig. 13.5a. The rest of the

a

Abd al-Gani al-Fakhri mosque, 1418. The probable source of Bourgoin's drawing (Tiling Search Web Site 2017, data186/P075)

b

Minbar from the same mosque, see (Abd-al-Wahhab 1946, Vol. 2, Fig. 151)

Fig. 13.6 Bourgoin's Plates 75 and 166

construction follows by making the heptagon circle-inscribed, so that calculating the radius of the circle determines the outline of the geometry. The result is Fig. 13.6a.

This approximate construction method contrasts with that of Bourgoin's Plate 166 which is exact due the star and the heptagon being such that all angles are a multiple of $\pi/14$, see Fig. 13.6b. Note that the polygon which was a regular hexagon in Bourgoin's original drawing of Plate 75 is now irregular and the point of three-fold symmetry is lost. An excellent modern photographic is available here so we can be sure that this pattern is correct; it actually comes from the same mosque in Cairo as Plate 75.

Verdict: We have no evidence that Bourgoin ever used a camera. In fact, the earliest example we know of a camera being used to record Islamic patterns is that of Edmund W. Smith (1894–1899) at Fatehpur Sikri. Hence Bourgoin's mistake is hardly unexpected.

13.4 An Egyptian Qur'ān

Many Qur'ān manuscripts have a front page with a geometric pattern. The one considered here is from the National Library of Cairo and is likely to be Mamluk period. In several cases, such illustrations for a Qur'ān manuscript contain problems from the point of view of a mathematical analysis; the example here is no exception.

A striking feature of this design is a decagonal area mainly in yellow. This area is well drawn according to the angles and lengths illustrated in Chap. 12. In contrast, the area outside these decagons is irregular in a way that is not forced by the topology of the design. Hence the solution is to re-draw the outer area using decagonal principles. This is illustrated with the computer graphic, Fig. 13.7b.

Verdict: The artist who drew this patterns seems to have had an off-day! Perhaps the master drew the decagons, but the apprentice drew the rest. Note that the 7-sided red polygons and the 12-sided blue polygons are not polygons found in other decagonal patterns.

a

Qur'ān from the National Library and Archives of Egypt. UNESCO, Memory of the World. Qur'ān dated to the Mamluk period (1250-1517)

b

Redrawn after analysis (Tiling Search Web Site 2017, data183/CD3)

Fig. 13.7 Photo of Qur'ān and computer graphic

13.5 Conclusions

We have seen in Chap. 9 that a large class of patterns can be fully determined. Examples given in this chapter show that many patterns can be indeterminate. In those cases in which a pattern is not fully determined, a complex analysis needs to be undertaken to give an acceptable result. Both mathematical and artistic aspects need to be considered. A more complex example is considered in Chap. 16.

Of course, in some cases, mistakes are made by the original craftsmen, or perhaps by poor restoration work.

Chapter 14
6-Fold Delights

In this chapter we look at patterns whose symmetry is either 632 (*p6*) or *632 (*p6m*), that is, there is a rotation of order 6. Hence such patterns were not included in Chaps. 9, 10 or 12. We first consider the impact of earlier mathematical designs on those of Islam.

14.1 Roman Influences

Figure 14.1a is interesting since it might be thought that the use of stars in geometric design was first used in Islamic times. This pattern is in a Roman mosaic from Thurburbo Majus (modern Tunisia), although the star contains two triangles of scroll work, perhaps making it less obvious. Of course, in mathematical terms, the design is very simple.

Figure 14.1b is a classical tessellation known to the Greeks which appears in Roman mosaics in Ostia and Pompeii in Italy. It also appears in a mosque in Turkey. Figure 14.1c is in a Roman mosaic from Sfax (again modern Tunisia). The same pattern is found in a Persian manuscript and a tile from Pakistan.

Figure 14.2a is a Byzantine style pattern from St Maria in Cosmedin, Rome. The design has one degree of freedom, the width of the white banding.

Figure 14.2b is a pattern drawn by James Wild in the 1840s which appears to be from al-Fadawiyya Mosque in Cairo. This design does not have a mirror-line and therefore has symmetry 632 (*p6*).

Figure 14.2c is a very similar design to Fig. 14.2b but is a Roman mosaic from Aquileia in Italy. The original has ornamentation not shown here. This also has symmetry 632 (*p6*).

B. Wichmann and D. Wade, *Islamic Design: A Mathematical Approach*, Mathematics and the Built Environment 2,
https://doi.org/10.1007/978-3-319-69977-6_14

a
Roman mosaic, Thurburbo Majus (Balmelle 2002, Plate 209 (c))

b
Roman mosaic, Ostia and Pompeii (Pajares-Ayuela 2002, Figs. 3-18f and 3.22b)

c
Roman mosaic, Sfax (Balmelle 2002, Pl 210 (c)), also woven textile (Victoria and Albert Museum collections website 2017, 109136)

Fig. 14.1 Roman or Byzantine patterns

a
Mosaic floor, St Maria in Cosmedin, Rome, 6th-12th centuries (Hessemer 1990, p 76)

b
al-Fadawiyya, Cairo (Victoria and Albert Museum collections website 2017, 1178103)

c
Roman mosaic, Aquileia (Balmelle 2002, Plate 205 (d))

Fig. 14.2 Roman or Byzantine style

a

Abd al-Samad Complex, Natanz, 1304 (Broug 2008, p 46). Also in Cairo and India, Fig. 6.2

b

Imam al-Shafi'i, Cairo, 1211 (Bourgoin 1879, Plate 2)

c

North Africa, 15th century (Tiling Search Web Site 2017, data18/W59) and Fatehpur Sikri, The Large Baths

Fig. 14.3 Some simple designs

14.2 Simple Islamic Designs

Here we illustrate three designs without stars.

Figure 14.3a is found in Iran, Cairo, India and Central Asia. This pattern has interlacing with a 6-way node. The thickness of the two waists on the 10-sided polygon can be varied. Only one example from Tosh Hovli Palace (Khiva) has true interlacing which is poorly rendered; in fact, the interlacing cannot be produced in a consistent fashion. The graphic here is consistent, but the 6-way intersection is not correct. The version from Cairo is very slightly different in that the 6-way intersection does not appear but is replaced by a small triangle, thus avoiding the problem.

Figure 14.3b is from Imam al-Shafi'i, in Cairo. The relative size of the two hexagons can be varied.

Figure 14.3c is from North Africa. Again, the pattern can be varied by changing the relative sizes of the diamonds and hexagons.

14.3 Simple Islamic Designs Without Stars

In this section we illustrate three designs with symmetry 632 (*p6*), and hence without a mirror line.

Figure 14.4a has no stars but the same outline is produced with diamonds. The pattern is from Imamzada Yahya in Iran and the colour has been added. Note that

a

Imamzadeh Yahya, Varamin (Pope 1965, Photo 182)

b

Great Mosque, Damascus (Tiling Search Web Site 2017, data174/SYR0130)

c

Uzghend, South Mausoleum. Carved portal plaster, 1186 (Tiling Search Web Site 2017, data185/AJL1)

Fig. 14.4 Simple designs without stars

this is not edge-to-edge. This pattern can easily be produced by positioning all the lines on a triangular grid.

Figure 14.4b is from the Great Mosque in Damascus. This is constructed from stone inlay. Given that the shorter edge of the black triangle has the same length as the side of the hexagon, the pattern is completely determined.

Figure 14.4c is from the South Mausoleum in Uzghend, Turkestan. It has a curious 'Z' shape similar to Fig. 14.4b. The pattern is also completely determined.

The patterns in Fig. 14.4 are all completely determined.

14.4 Simple Islamic Designs with a Single Rosette

Figure 14.5a is a well-known pattern (Bourgoin 1879, Plate 21) and has been widely copied from Bourgoin. It is thought that the version from Morocco has also been copied from Bourgoin, see page 183. The graphic presents a problem since there are two parameters to determine.

A photo of this pattern by Bernard O'Kane is now on the Internet and he writes: The wakala (urban warehouse with dwellings) of the Mamluk Sultan Qaytbay was built in 1477. He built two; this is the one behind al-Azhar mosque. It is carved in stone (without colour), but with interlacing. The proportions have been adjusted to reflect this version but with colouring from Morocco. The lines of the pattern could follow a larger version of the 6-pointed star, but this is not the case here. Hence the

a

Wikala of Sultan Qaytbay, Cairo, 1477 (Alamy 2017, Ref BX29GN)

b

al-Nasir Muhammad Mosque, Cairo (Broug 2008). Other sites in Turkey Syria and Morocco (Bourgoin 1879, Plate 69)

c

Eastern tower, Kharqan, Iran, 1067-8 (Lee 1987, Fig. 10). Widespread pattern. (Here the colour is from a Turkish example.)

Fig. 14.5 Simple designs with a single rosette

diamonds are further away from the 6-pointed star than the black kite. With these proportions, the pentagon is nearer to being regular than would otherwise be the case.

Figure 14.5b is also a well-known pattern of (Bourgoin 1879, Plate 69). Unlike (a) this is a common pattern and appears at least twice in the Great Mosque in Damascus, in Cairo, Morocco and Turkey. The star has a vertex angle of 60°, which determines the surrounding kites. Assuming the conventionally-shaped petals are standard (the four edges further from the main star being of equal length), then the pattern is very nearly completely determined. The only uncertainty is the angle at the far end of these petals. All the other angles are multiples of 15°, so we take the angle between the adjoining petals as 30°.

Figure 14.5c is a widespread of Islamic star pattern, in various materials, from Morocco to India. It has given rise to numerous variations. The vertex angle of the 12-pointed star is 90°, which determines the adjoining kites, thus the pattern is completely determined. This leaves a small triangle at the point of 3-fold symmetry.

Note that all three of these designs have non-standard 'petals' (not 6-sided with four lengths the same).

a

Qaytbay, Cairo, 1471 (Tiling Search Web Site 2017, data12/P110). Also in Turkey (Bourgoin 1879, Plate 110)

b

al-Mu'ayyad, Cairo, 1420 (Patterns in Islamic Art web site 2017, EGY 1216)

c

Qur'ān in British Library, 1568, also in Morocco (Tiling Search Web Site 2017, data177/J43C). Cannot be constructed with ruler and compass

Fig. 14.6 Designs with two rosettes

14.5 Designs with Two Rosettes

Figure 14.6a is another common pattern is (Bourgoin 1879, Plate 110). This is not an easy pattern to draw since the vertex angle of the two stars is not obvious; indeed they can be varied slightly. On the other hand, we assume that the octagon is regular.

The construction depends upon three circles, around the two stars and the octagon. The lengths of the radii give the edge lengths of the sides of the triangle joining the centres of the two stars and the octagon. Also, the edge lines of the petals of the two stars can be collinear. This restricts the possible vertex angle of the main star; values of 75° and 80° are possible, but 75° gives a better shape to the single non-rosette polygon. Making the final decision to use standard petals completes the design process.

Construction can now proceed. The main star is determined by taking the edge length of the star as unity. One can now see that the angle between adjoint petals is 45° which is exactly the angle needed between the octagon edges of the outer edge of the main petal. This confirms that 75° is the correct angle of the vertex of the main star. If the circum-radii of the two rosettes are R and r, and that of the octagon p, then:

$$tan(30) = (r + p)/(\mathrm{R} + p)$$

$$sin(30) = (r + p)/(r + \mathrm{R})$$

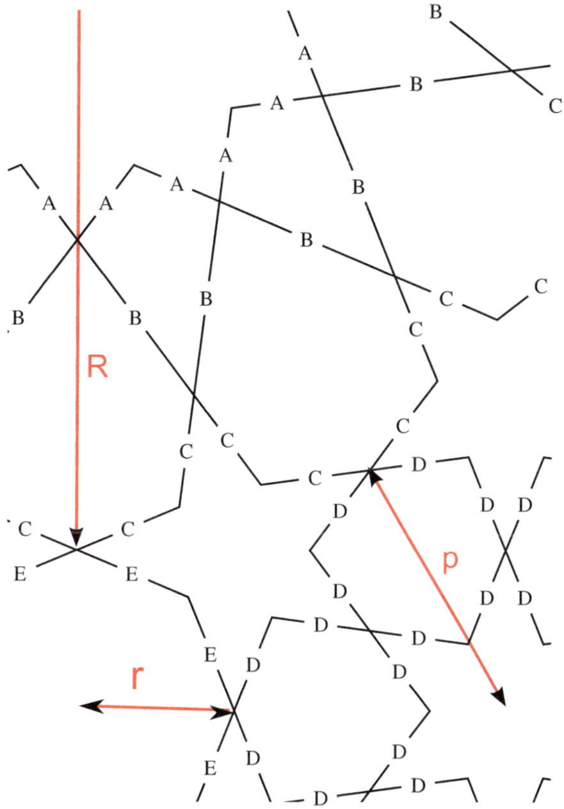

Fig. 14.7 Construction of Fig. 14.6a; the angles are multiples of 15°

from which all the values can be determined. This construction can be seen in Fig. 14.7. The pattern shows the difficulty of choosing the correct vertex angle for patterns when the constraints do not given a specific value.

Figure 14.6b has a star with nine petals which implies that the pattern cannot be constructed with ruler and compass alone (Jagy 2017). The clearest version of this pattern appears to be a panel of a minbar from the Mamluk period. The woodwork produced during this period is of the highest quality and therefore can be taken as the basis for an analysis. A modern photograph is shown in Fig. 14.8.

The related drawing by (Bourgoin 1879, Plate 120) of this pattern shows a small interlace discontinuity along the longest edges of the two 'petals' which are not part of a rosette. Such small imperfections would be acceptable in Islamic patterns, but we can avoid that in our case. We can take the two non-rosette petals as being the same as for the 9-pointed star rosettes.

The construction is shown in Fig. 14.9. The angle marked Y is 30°. Hence to maintain an elegant 5-pointed star, albeit not regular, would be to have the other angles marked X as 30° also. Noting that the petals around the 9-pointed star are rotated by 40°, if those petals had parallel sides, the angle between the petals would

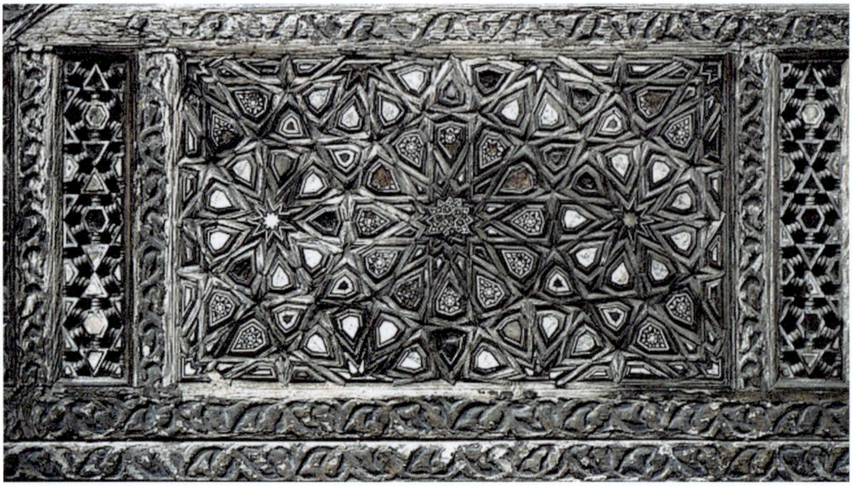

Fig. 14.8 Mamluk woodwork from a minbar in Mosque of al-Mu'ayyad, Cairo (Patterns in Islamic Art web site 2017, EGY 1216)

Fig. 14.9 Construction of Fig. 14.6b

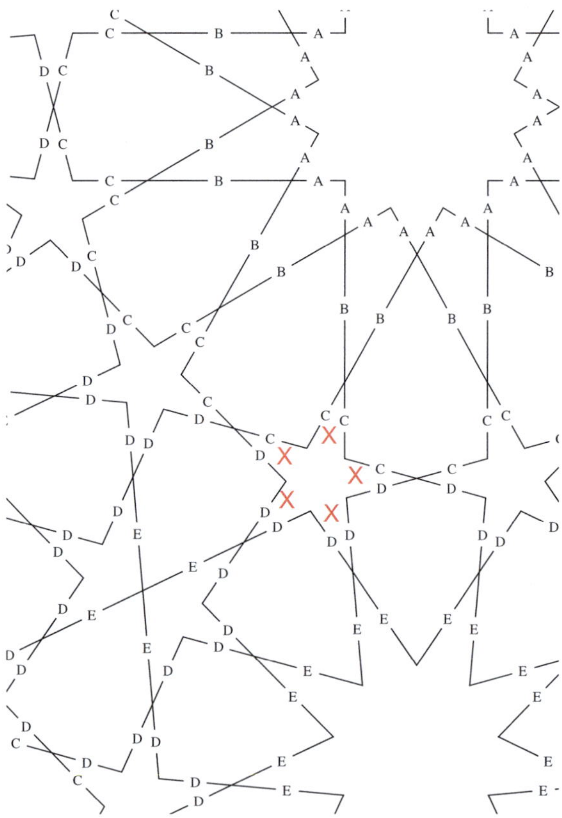

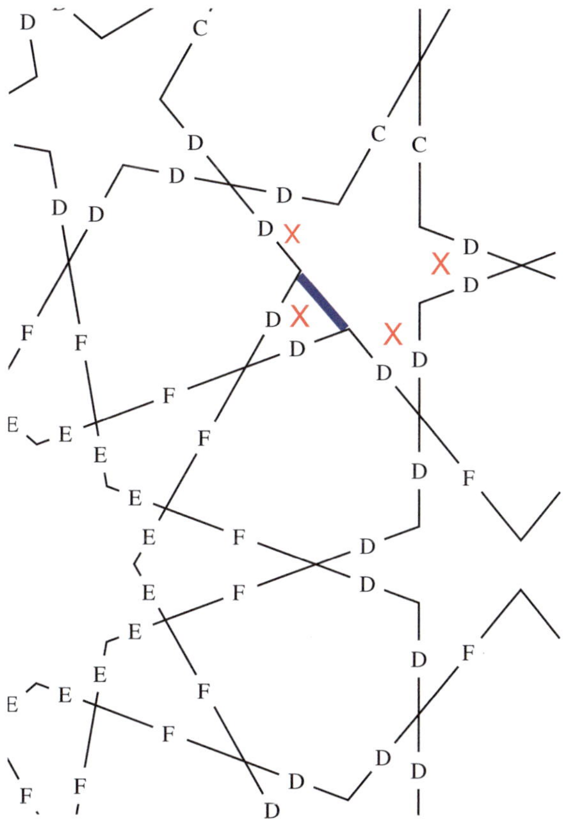

Fig. 14.10 Construction of Fig. 14.6c

be 40°. Hence making the sides diverge by 5° would give the angles we wish. This allows all the edge lengths as well as the angles to be determined.

Figure 14.6c is a member of the small group of related patterns which includes Fig. 14.6b. This group has 9-pointed and 12-pointed rosettes in a similar configuration but differ around the area of the two non-rosette petals. In pattern Fig 14.6c, the petals are not joined at a point; there is a small gap. In others, there is actually an overlap.

In this variant, the 12-pointed petals are not standard, but the sides are extended slightly. The 9-pointed rosette has standard petals with parallel sides and also has kites (unlike Fig. 14.6b).

The construction is illustrated in Fig. 14.10. The outer lines of adjacent petals are collinear (shown in blue) which implies all the angles X are 40°. All the edges either side of these angles have the same length (marked as D). This implies that the length C is determined.

a

Morocco, dated 1940 (due to the date, likely to be a modern copy), (Paccard 1980, vol. 1, p. 456); Gayer-Anderson Museum, Cairo

b

Alhambra, Salon de Comares, 14th century (Fernandez-Puertas 1999, p. 423, Fig. 224), Also in Morocco

c

Zawiyat al-Abbar, Cairo, 1283 (Creswell 1952/59, vol. 2, Plate 61c), Also in India, Bukhara and Iran

Fig. 14.11 Some complex designs

14.6 Some More Complex Designs

We consider three more examples of patterns with 6-fold symmetry which are slightly more complex than the previous examples.

Figure 14.11a is a striking design from Cairo, which was probably copied by Moroccan craftsmen. This pattern is usually referred to as (Bourgoin 1879, Plate 96) (also (Bourgoin 1879, Pl.VIII) round the fountain in men's reception room at Kritiya House, Cairo). The analysis and construction of this pattern is straightforward. To have parallel-sided 'petals', the vertex angle of the main star is 60°. This determines the surrounding kites. Taking the kites around the other stars as being congruent, then it is clear that the four outer edges of the main 'petals' all have the same length as the edges of the main star. Hence the edges of the 3-lobed red polygons are all of the same length.

Figure 14.11b appears in the Alhambra. The repeated motif of this graphic is taken from the large, complex ceramic mosaic dados in the central throne alcove in the north wall of the Salon de Comares. This dado includes a number of different 12-fold motifs, which are used individually to form repeating patterns in these graphics. Two such motifs appear together with other motifs are from the same dado, in Alhambra, Salon de Comares. The present pattern is common in Morocco, and this type of 12-fold motif seems characteristic of western Islamic geometric ornament. Like Fig. 14.11a, the main star has a vertex angle of 60°. Note that there is a bounding regular dodecagon around the 10-sided polygons outside the 'petals'. This dodecagon determines the rest of the pattern including the hexagons at the points of 3-fold symmetry. The angle of the cross-overs seems to be 60° matching that of the main star. The length of the sides of the 'petals' and the proportion of the cross-overs to the width of the petal is not determined and can only be by measurement of the several physical examples of this pattern.

Figure 14.11c is an interesting pattern which is widespread in Egypt, Central Asia and India, 13th to 16th centuries; not present as far as we know in the Maghreb. The analysis depends upon the dodecagon surrounding the main 12-pointed star. This implies that the largest angle of the kite around the 9-pointed star is 150°. Following the line from the outer edges of the same kite to the main star gives one line of the triangle with the opposite vertex at the centre of the main star (see the two red lines Fig. 14.12 bounded by the third side of length $A + B$). The angles of this triangle show that the vertex angle of the main star is 100°. The line with arrows has similar lines around the main star, also making a dodecagon. Taking the edge length of the main star A as unity, all the edges can now be computed. The formula for P uses the triangle formed by the red line with the arrows and the edge of length B whose angles are 75°, 55° and 50°.

Formula
$A = 1.0$
$B = A * \sin(65)/\sin(35)$
$P = B * \sin(50)/\sin(75)$
$C = P * \sin(70)/\sin(55)$
$D = C * \sin(35)/\sin(75)$

Note that the kite marked Z is the same as the kites around the 9-pointed star. The pentagons are not regular, but nearly so.

14.7 A Motif with Two Symmetries

We consider here a motif that has been used in Morocco in two different forms having $*442$ (*p4m*) and $*632$ (*p6m*) symmetries. There is a modern copy of the second form as an ornamental door of the Hassan II Mosque in Casablanca. (This modern version has a small interlacing error.)

The motif has a 24-pointed star; of course the number of points must be multiples of 4 and 6. We first consider the motif itself. Fig. 14.14 shows an expanded view of the $*632$ (*p6m*) version to illustrate the construction.

Fig. 14.12 Construction of Fig. 14.11c

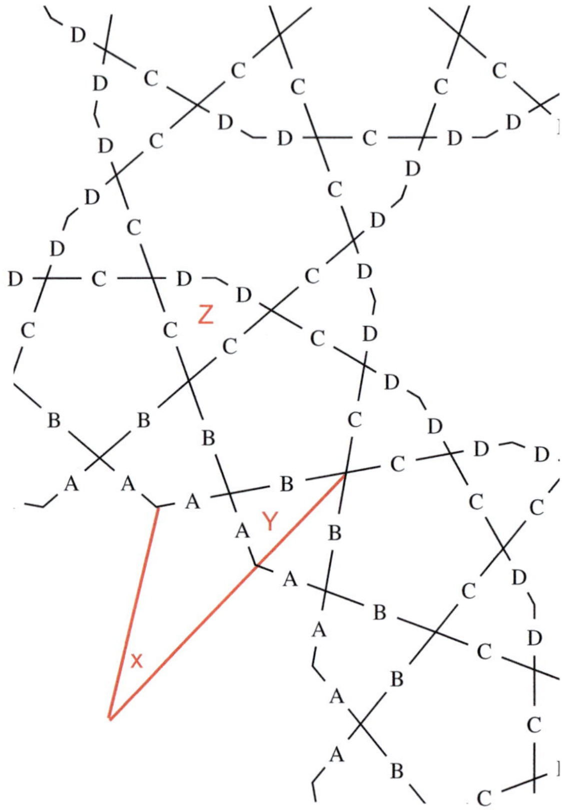

a

From Fez, exact location not known
*632 (*p6m*)

b

Tomb of Moulay Ishmail, Meknes, 1732, see (Patterns in Islamic Art web site 2017, MOR 2443X), *442 (*p4m*)

Fig. 14.13 Motif with two symmetries

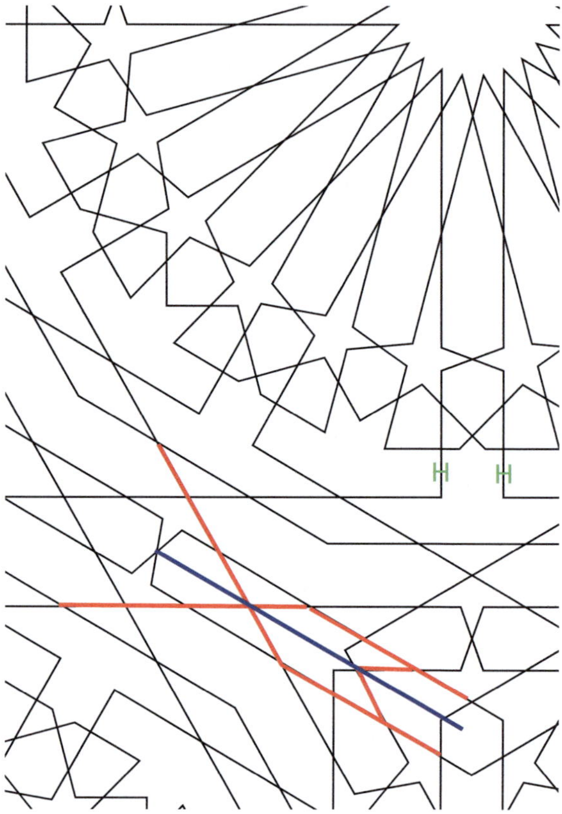

Fig. 14.14 Construction of motif in Fig. 14.13

The vertex angle of the main star must be 30° to ensure the petals are parallel sided. If the width of the petals is W, then the edge length of the hexagon is $C = W/(2 * \cos(30))$. All the individual line segments shown in red are of the same length $B = W$. This implies that the length of the blue line is determined, which therefore gives the size of the motif. However the motif is not entirely fixed since the length marked H can be varied.

Returning to the $*442$ (*p4m*) variant, there is a square shaped motif to fill in around the corners. This motif is essentially the pattern in Fig. 14.11b.

Chapter 15
Two-Level Patterns

Two-level patterns are simply defined as patterns within patterns (Wichmann 2001). In a detailed study, Peter Cromwell (2016) lists forty examples; but in the first part of this chapter we consider just the simpler ones. In the second part, we look at more complex examples and contrast the simple approach with the methods used by the designers in Iran.

This topic has been controversial due to a publication in *Science* (Lu and Steinhardt 2007). We agree with Cromwell that there is no evidence of quasi-periodic tilings, merely two-level patterns. Even three-level patterns would provide some difficulty since the ratio between the smallest tile and the uppermost pattern would be too large to accommodate in many contexts.

15.1 An Iranian Example

As a simple example we consider a two-level pattern from Iran shown in both a photo and a computer drawing in Fig. 15.1. It is important to compare the two, since they differ in small details. Some of the khatems have central circles which are omitted in the computer version. Also, the colouring is inconsistent in the photo.

There are at least two significant differences: the khatems which cross the upper-level boundaries are coloured in either blue or stone (divided into two or four parts); and secondly, the boundary lines in the photo are quite wide but do not overlap the pattern (hence the alignment between the parts is not exact). In contrast, in the computer drawing, the upper-level lines are added on top of the lower-level drawing and hence hide part of the lower level design. For this reason the upper-level boundary lines are narrower. The lower-level pattern lines are shown in white which gives a clearer result.

B. Wichmann and D. Wade, *Islamic Design: A Mathematical Approach*, Mathematics and the Built Environment 2,
https://doi.org/10.1007/978-3-319-69977-6_15

a

Pattern dated 1840, modern photo (Patterns in Islamic Art web site 2017, IRA 1617)

b

Computer graphic using standard tiles overlaid with lines to show the upper pattern

Fig. 15.1 Seyyed mosque, Isfahan

Of course this design, at both levels, is a member of the set of patterns considered in Chap. 9. Although the drawing differs from the photo, it captures the design accurately (apart from the differences noted above).

15.2 An Indian Example

Our second example is from India which is another significant source of two-level patterns Fig. 15.2.

For the screen itself, we can do no better than copy the museum details:

> Jalis, or pierced screens, were used extensively in Indian architecture as windows, room dividers, and railings. In the course of the day, the movement of their patterns in silhouette across the floor would enhance the pleasure of their intricate geometry. This jali, one of a pair, would have formed part of a series of windows set in an outside wall, as suggested by the weathering on one side. They are attributed to the reign of the Mughal emperor Akbar (r. 1550-1605), when red sandstone was the favoured building material.

a

The Metropolitan Museum of Art (1993.67.2)

b

Computer version using two widths of lines

Fig. 15.2 Pierced window screen

There are some points to note: The khatems are divided into four which is likely to be undertaken for structural integrity. (There is another version of this pattern in Salim Chishti's Tomb, at Fatehpur Sikri. This second version divides the khatems differently, perhaps in a somewhat less elegant fashion.) The computer graphic is produced by overlaying the upper-level on top of the lower-level pattern. In consequence, the lower-level crosses the upper-level lines directly, which is not the case in the actual Jali screen. We will return to this point later.

15.3 A Decagonal Example

This more complex decagonal pattern in Fig. 15.3 shows only the computer graphic.

The blue line giving the upper-level pattern is narrow so that as little as possible of the lower pattern is obscured. Considering symmetry of the tiles in the upper-level pattern:

- The main star itself has symmetry *10• (*d10*) and surprisingly, the constituent tiles retain this symmetry.
- All the kites are identical and have a mirror-line as the only symmetry.

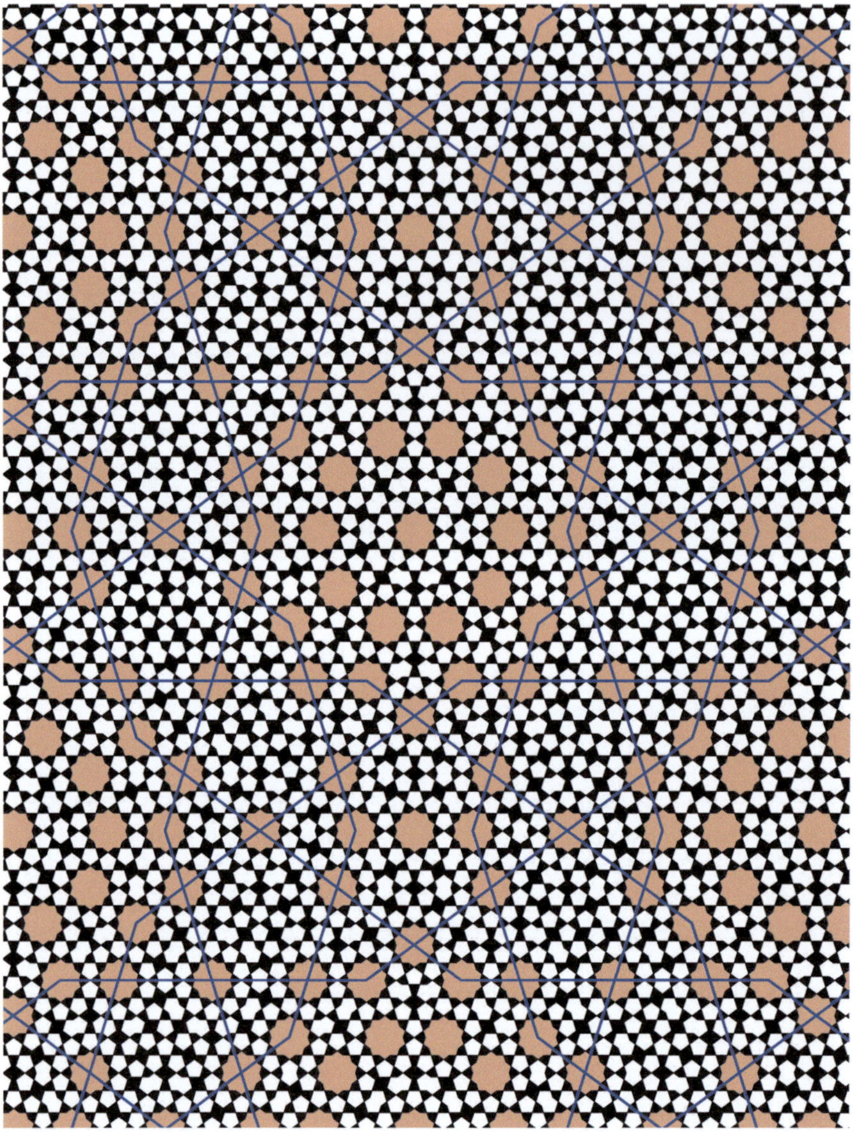

Fig. 15.3 Jame mosque, Isfahan (Patterns in Islamic Art web site 2017, IRA 0601). The upper-level pattern in blue is Fig. 12.1a

- The pentagons are also identical (suitably rotated), but have no symmetry. (Note that the pentagons could have been replaced by the motif in Fig. 12.2a and then have symmetry 5• (*c5*)).
- The 8-sided figure has *2• (*d2*) symmetry as a single polygon and also with the constituent tiles.

15.4 The Topkapı Scroll

Our last example is from the Topkapı scroll (Necipoğlu 1995, p. 300, Fig. 28). The scroll drawing provides excellent detail but is without coloured tiles. We show the original Topkapı *girih* drawing in Fig. 15.4; an analysis of this appears in (Cromwell 2016, Fig. 18).

The drawing might appear to have a confusing number of lines, but it is very accurately drawn and shows all the necessary detail. The upper-level pattern is shown with long red lines, so there is a star with kites centered on the top left corner. The lower-level pattern is shown with short black lines; there is a snake of 5-pointed stars going upwards from the middle of the bottom edge.

What are the other red lines which are faint and shorter than those showing the upper-level pattern? According to Daud Sutton (2007, p. 34), this is *Umm al-Girih* which is Persian for mother of pearl (knots). The use of these lines should be clear since the pattern has all the essential elements related to a mesh of decagons which meet each other. The drawing can therefore be produced by first producing the

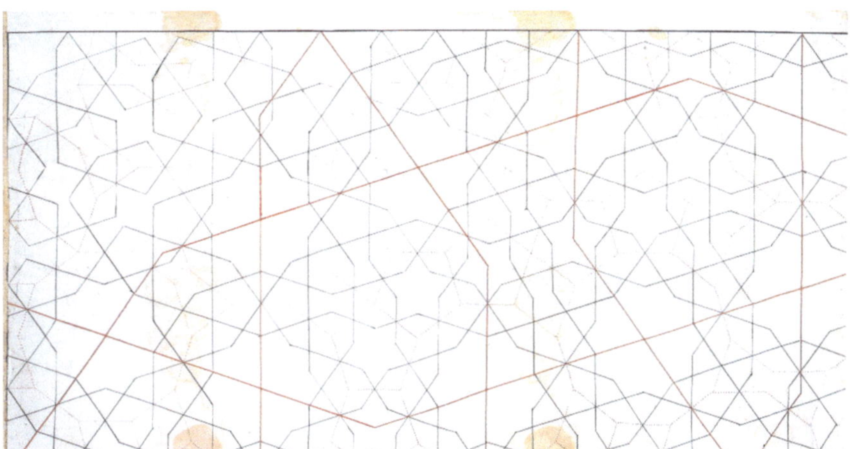

Fig. 15.4 Topkapı scroll—original

Fig. 15.5 Topkapı scroll—graphic. Colour added to show the structure. No examples is known of this pattern apart from the scroll. The upper level pattern is shown with blue lines

decagons which then show how all the remaining lines can be drawn. Some of those lines produce pentagons which are bigger than those of the lower-level pattern.

The main star this time has a vertex angle of 72°, instead of 108° as in the previous figure.

- The main star itself has symmetry *10• (*d10*), and again, the constituent tiles retains this symmetry.
- The kites round the main star are all identical and have one mirror-line.
- The pentagons are also all identical and have symmetry *5• (*d5*).
- The 6-sided bowtie figure has *2• (*d2*) symmetry as a single polygon and also with the constituent tiles.
- The other 6-sided figure has a single mirror-line as a single polygon and also with the constituent tiles.

This figure therefore has a remarkable property: each individual polygon at the upper-level has the same symmetry when considering the constituent tiles. (Actually, Fig. 15.1 has the same property.)

Finally, we can compute the ratio in scale between the two levels of the patterns:

Figure	Ratio
Fig. 15.1	4.83
Fig. 15.2	4.83
Fig. 15.3	8.47
Fig. 15.5	5.24

We can now return to the problem of having three-level patterns. From the above table this would appear to indicate a ratio between the smallest pattern and the largest of over twenty. However, even with Fig. 15.3, the ratio of 8 provides some difficulty for the eye to see both levels.

15.5 A Complex Issue to Resolve

The issue which has been ignored earlier is that of the upper-level pattern hiding some of the lower-level pattern. We provide a solution to that problem, which is then applied to two demanding examples.

With the graphic in Fig. 15.3, we did not show a photograph for two reasons for this: firstly, as noted above the ratio is large which implies the detail is small secondly, the artefact seems to be poorly produced or has been damaged. However, we need to consider how such a pattern is produced if the blue lines were thicker.

The problem is that if the blue lines are merely put on top of the existing graphic, some of the underlying pattern will be obscured. So perhaps the pieces of the underlying pattern can be cut up along the blue lines and then reassembled along the edges of wider blue lines.

We consider this issue in more detail with a specific related example starting from Fig. 15.6. Here, the upper-level pattern is in black, but uses the same narrow width as the lower-level pattern in red. The black pattern is the same as in Fig. 15.3, but the lower-level pattern is simpler. Figure 15.7 shows the consequences of drawing a thick line for the upper-level pattern.

Fig. 15.6 Basic two-level pattern

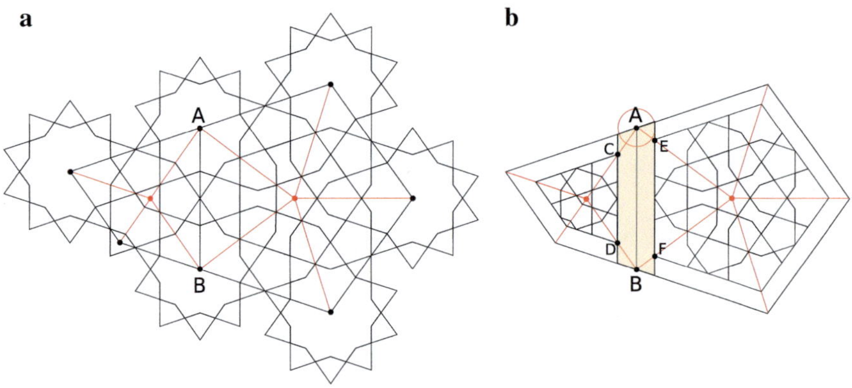

Fig. 15.7 **a** shows a small area with a pentagon and a kite, while **b** shows the same area redrawn for a two-level pattern

The important point to note that the line EF is not the same length as CD so that the kite needs to be drawn to a *different scale*. These scales depend upon the width chosen for the upper-level pattern. Of course, we have chosen a wide line AB in Fig. 15.7 to illustrate the issue.

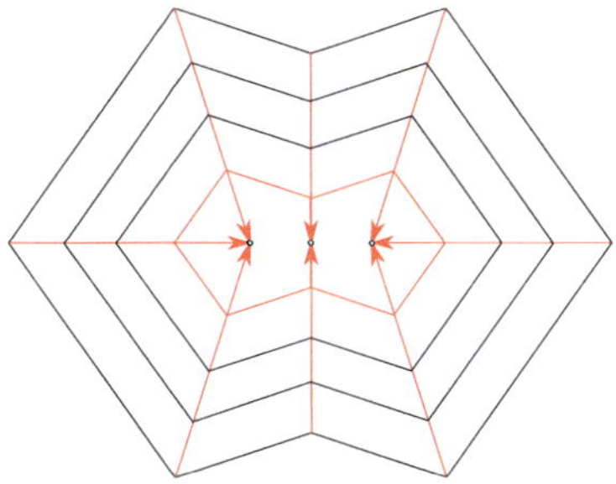

Fig. 15.8 Reduction in size of a twinned-pentagon shape by successively thicker banding

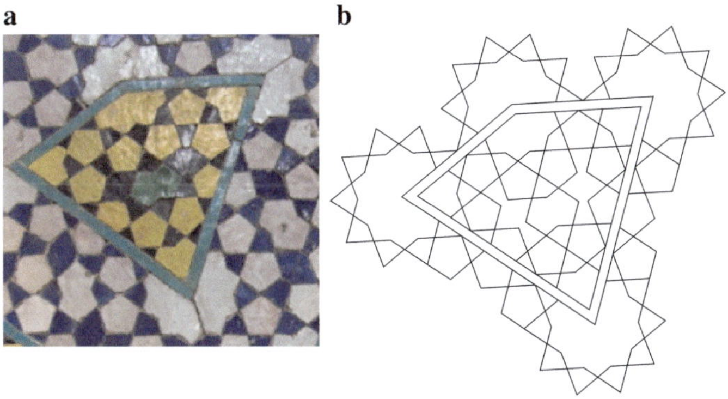

Fig. 15.9 Detail of the kite in a two-level pattern with a diagram of its geometry, and showing the difference in scale between kite and surround

If the internal angle bisectors of a polygon do not meet at a single point, then the polygon will not retain its exact shape when the presence of upper-level banding produces shrinking. This is shown clearly by the "twinned-pentagon" shape in Fig. 15.8. Here the outer edges represent the original positions of lines in the upper-level pattern (half the shape is visible midway along each border of the black pattern in Fig. 15.6). Equal thicknesses of banding on each edge reduces its original size, shown successively by the smaller polygons within the outer edges, but the initial shape gradually changes, until the outline in red is equivalent to a pair of small pentagons back to back.

This whole process is now quite complex, so why consider this at all? The answer is that actual two-level patterns which we shall describe later show these issues, so if they are to be drawn faithfully, we have no option. However, due to the complexity, we do not undertake the complete calculation necessary to give mathematical precision, but use a computer-drawing package to obtain the desired effect.

To illustrate the need for the change of scale in a real Islamic two-level pattern, we take a decagonal pattern from Isfahan, from the Madar-i Shah Madrasah. In Fig. 15.9 a lower-level pattern is shown as the surround, while the kite has a broad blue line around it. (We call the drawing within the kite a *module*.) The lack of continuity between the inside of the kite and the surround is easily seen from the computer drawing. This also occurs in the Isfahan pattern, but is less clear due to the colouring.

Note that the small change in shape of the polygon in Fig. 15.8 would be very tricky to undertake with mathematical precision. Indeed, it is unclear what the effect this would have on the constituent lower-level pattern.

15.6 A Complex Pattern: Darb-I Imam Shrine

We can now apply the method of handling modules given above to a complete pattern.

The drawing in Fig. 15.10 applies the methods given in the last section in the manner specified. The upper-level pattern is that shown in Figs. 15.3 and 15.6. The black lines are wide enough to show the corrections needed to the scale of the individual modules. When a white star is broken into two or four parts, the result appears to be a mis-shapen star; this is also a feature of the original Isfahan mosaic (see Fig. 15.9a). The original also shows slight inaccuracies in the layout of the black banding of the upper-level pattern. The orientation of the patterns of tiles inside the pentagonal modules is variable in the original, but these are not accurately copied in the drawing above.

15.7 A Complex Pattern: Seyyed Mosque

A complex pattern is the high point of Islamic design (Fig. 15.11). It is ‘a challenge to produce correctly’ when using a computer. The lower-level design in the central star has been omitted, since the original mosaic does not contain any consistent or symmetric arrangement of tiles in this region.

The basic underlying pattern is quite simple and can be seen in Fig. 12.11. With the main star omitted, the lower-level modules needed for the banding are the kite, the petal, the 5-pointed star and the 6-sided arrow-head shape.

The pattern is an example of banding in which the greatly thickened ‘lines’ of the pattern in Fig. 12.11 themselves form a similar decagonal pattern.

The original mosaic at the Seyyed Mosque (dated 1840), from which Fig. 15.11 has been taken, has very thin strips of secondary banding between the lower-level pattern in the bands of the upper level pattern, and the in the various modules between the bands of the upper pattern. However, because of the very acute angles of the small kites, peripheral 5-pointed stars and arrow-heads, the lower-level pattern inside these is very obviously at a much smaller scale than that in the upper-level banding.

As initially laid out for Fig. 15.11, these separate areas of lower-level patterns were drawn at the same scale, but shrinking them to fit the smaller spaces remaining, after secondary banding automatically lead to a reduction in scale, as is also necessarily the case in the original mosaic. The acute sectors of the pale blue 10-pointed stars

Fig. 15.10 Two-level tile mosaic from arch spandrels in the Darb-i Imam shrine at Isfahan

Fig. 15.11 A two-level design based on a mosaic on spandrels over an arch at the Seyyed Mosque in Isfahan

Fig. 15.12 Seyyed Mosque in Isfahan showing a detail enlarged from Fig. 15.11

occupying the acute angles of the peripheral 5-pointed stars and the small kites in most cases appear to have been squeezed out of the rest of the 10-pointed star, although in some cases the mosaicist seems to have taken pains to reduce this effect. The most obvious evidence of the reduction in scale in certain areas of the lower-level pattern is in a comparison of the sizes of the pentagonal tiles each side of the secondary banding. This is very clearly shown on magnification of the drawing in Fig. 15.12, but it is also quite obvious on the original mosaic.

Theoretically in a two-level mosaic of this type the location of the secondary banding initially acts as a local mirror axis to the pattern on both sides. After the

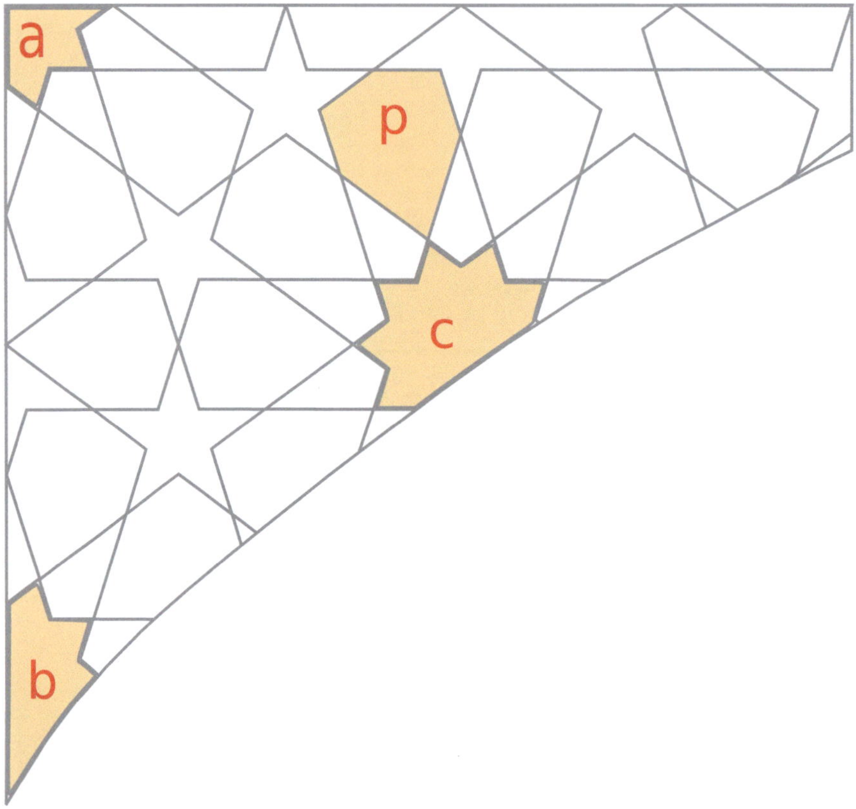

Fig. 15.13 The scheme of the upper-level pattern on the left spandrel of the Seyyed mosque. The locations of the central stars a, b and c are marked

introduction of secondary banding has led to scale reduction most authentic examples still attempt a kind of "topological" mirroring, in that the same sequences of shapes are retained each side of the secondary banding. However, if we attempt to complete a pattern inside the central 10-pointed star on this basis, we discover that it becomes extremely difficult if not impossible to arrange the usual set of decagonal tiles in a 10-fold pattern. Nowhere in the original mosaic have the artists managed to produce an arrangement of standard "decagonal" tiles which could be completed as a 10-fold design for the central star, and this fact probably reflects the extreme difficulty, and perhaps even the impossibility of doing so (areas *a*, *b* and *c* on Fig. 15.13). This appears to have been something of an embarrassment to the original artists, since they were obviously struggling to fill these areas with what seem almost random arrangements, with little or no symmetry. For this reason, the central is left 10-pointed star empty, rather than attempting to invent something.

Overall, this Seyyed Mosque spandrel pattern is a superbly conceived composition, achieved in a masterly manner—it does have at least one mistake—on the left hand spandrel part of the upper pattern banding was evidently allowed to become slightly distorted, with the result that the area of an adjacent petal was a little too large (area *p* on Fig. 15.13). Consequently the larger area had to be filled by adding an extra row of tiles along two of its edges. This seems to confirm that the upper-level banding was laid out first, then the various compartments filled in afterwards. Figure 15.13 shows that the spandrel pattern, if extended as a two-dimensional repeating pattern, would be slightly different from the one in Fig. 15.11.

It seems to have been a general principle in two-level patterns, in which the lines of the larger, or upper-level pattern were thickened as substantial bands, that any lower-level pattern drawn in the polygons or compartments of the large pattern should treat the edges of the thickened bands as its boundary, with the inevitable result that the isolated areas of lower-level pattern have to be drawn at different scales, depending on the chosen thickness of the upper-level banding, than would have been the case if the compartments had met one another along a geometrical line. This will always be so, if a discrete patterned area is separated from an adjacent, similarly patterned area by an appreciable thickness of either primary or secondary banding, and is a geometrical property, as demonstrated in Figs. 15.7 and 15.9. The fact that lower-level patterns have this relation to the thickened banding of an upper-level pattern is widely appreciated, but the fact of the necessary change of scale does not seem to have been mentioned, although it is perfectly visible on authentic two-level mosaics, as seen in Fig. 15.9.

To conclude, this chapter provides two means of drawing two-level patterns. The purely mathematical means of overlaying one pattern with another has clear limitations. The alternative means of following the methods used in actual artefacts is not practical if mathematical precision is required, but can be handled with computer-based drawing packages.

Chapter 16
Two Mamluk Masterpieces

The Mamluk dynasty which ruled Egypt and survived the Mongol depredations, produced some of the most impressive Islamic patterns particularly in wood on doors and minbars. Although our first example is in metal, it is an excellent example for which good photographs are available. It is Bourgoin's Plate 137 and is from the Madrasah and Khanqa al–Azahr Barquq, Cairo, Egypt. This was drawn by Bourgoin, but we redraw it using a detailed analysis. A photograph of the complete door is shown in Fig. 16.1.

The modern inscription is:.

B. Wichmann and D. Wade, *Islamic Design: A Mathematical Approach*, Mathematics and the Built Environment 2,
https://doi.org/10.1007/978-3-319-69977-6_16

Fig. 16.1 Door: Madrasah and Khanqa al-Zahir Barquq, Cairo

16.1 The Trigonometric Method

The first stage is to analyse the pattern to understand the essential details. There are two rosettes with 18 and 12 points. The regular 18-pointed star (like the 9-pointed star) cannot be drawn with rule and compass alone (Jagy 2017). The 12-pointed rosette has parallel-sided petals which implies that the vertex angle is 60°, but it is also standard with the outer four edges having the same length. The 18-pointed rosette is slightly convergent so the vertex angle is about 45°. A key aspect is the heptagons which should be drawn to be as regular as possible (although exact regularity is not possible). Extension of the sides of these heptagons then determines the relative sizes and shapes of the rosettes; but how to draw the heptagons at the start is largely left to the artistic sensibilities of the reader, rather than to any rigidly applied geometrical construction. Since the pattern is not completely determined, we can start by taking angles and lengths for the heptagon then work outwards to the rest of the pattern.

Figure 16.2 shows a photograph of the detail around the heptagon. The heptagon insert is clearly poorly produced, which might be due to later restoration. Also, the

Fig. 16.2 Door: Detail around heptagon

'lines' of the pattern are quite wide which makes it difficult to measure the angles precisely. Note that the outer edges of adjacent petals for the 18-pointed rosette are not quite collinear.

We can produce an outline of the required pattern from which we can calculate the lengths and angles (Fig. 16.3).

The calculate the edges of the heptagon we take L as unity and apply the sine formulae repeatedly:

Formula
$L = 1.0$
$M = L * \sin(105)/\sin(37.5)$
$R = L * \sin(50)/\cos(70)$
$T = M * \cos(27.5)/\sin(130)$
$U = M * \sin(22.5)/\sin(130)$

From which the value of S follows.

We now consider the triangle formed by the blue line of length M and the extended edges of E and F (we denote the extended edges as V and Q respectively). Since the small angle is 20°, the sine formulae give:

Formula
$V = M * \sin(82.5)/\sin(20)$
$Q = M * \sin(77.5)/\sin(20)$

The two distances marked Y are clearly the same, so $Y = 2 * E * \sin(65)$. Now consider the very thin triangle whose angles are 5°, 20° and 155°. Applying the sine rule we have $Y/\sin(20) = Z/\sin(155)$, where $Z + E = V$. Eliminating Y, we have $E/Z = \sin(20)/(2 * \sin(155) * \sin(65))$. This gives us E.

To calculate the value of F, consider again the very thin triangle whose angles are 5°, 20° and 155°. The shortest edge is $Y * \sin(5)/\sin(20)$. Hence $P = Q - Y * \sin(5)/\sin(20)$ for the remaining length of the extension of F (shown as a dashed red line). Considering the triangle of sides P and N (the lower on in the figure), we have $N = P * \sin(60)/\sin(45)$. Hence we have $F = N * \sin(15)/\sin(60)$.

This effectively completes the calculation of the pattern based on our analysis. Although the Arabs as part of the House of Wisdom in the 10th century knew the sine rule, we do not think this calculation was performed to construct this pattern. As an alternative, we present another method of constructing this pattern based upon ruler and compass rather than trigonometry. This 'manual' method depends upon making adjustments to give an acceptable result.

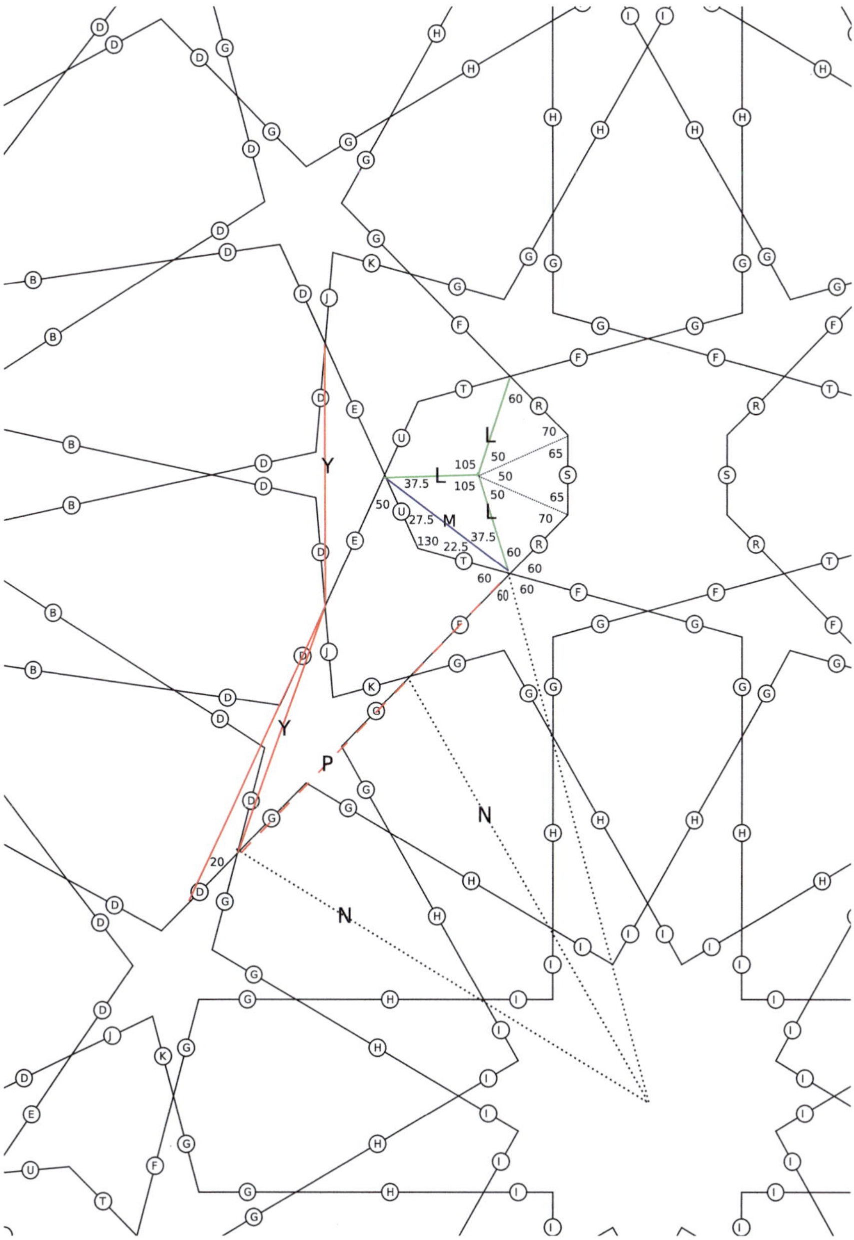

Fig. 16.3 Lengths and angles around heptagon

16.2 The Manual Method

We start with the basic layout Fig. 16.4, and then insert the 'heptagon'. This shows a basic rectangle—points A, B, A, B—which will fill any required pattern area by mirroring through its four edges. Note that GF = GC∗ tan(15) = GA∗ tan(10) giving the relative size of the two rosettes. The 18-pointed rosettes are centred on points A, and 12-pointed rosettes on points C. Hence both rosettes are surrounded by regular polygons sharing a common edge FGF. This construction is called *polygons-in-contacts* (Lee 1987, p. 189). Around point C on the left hand side we draw radii at equal angles of 15°. From upper left point A draw two radii at angles of 10° to the vertical line between A and C. Points F are the intersections of radii from A and C,

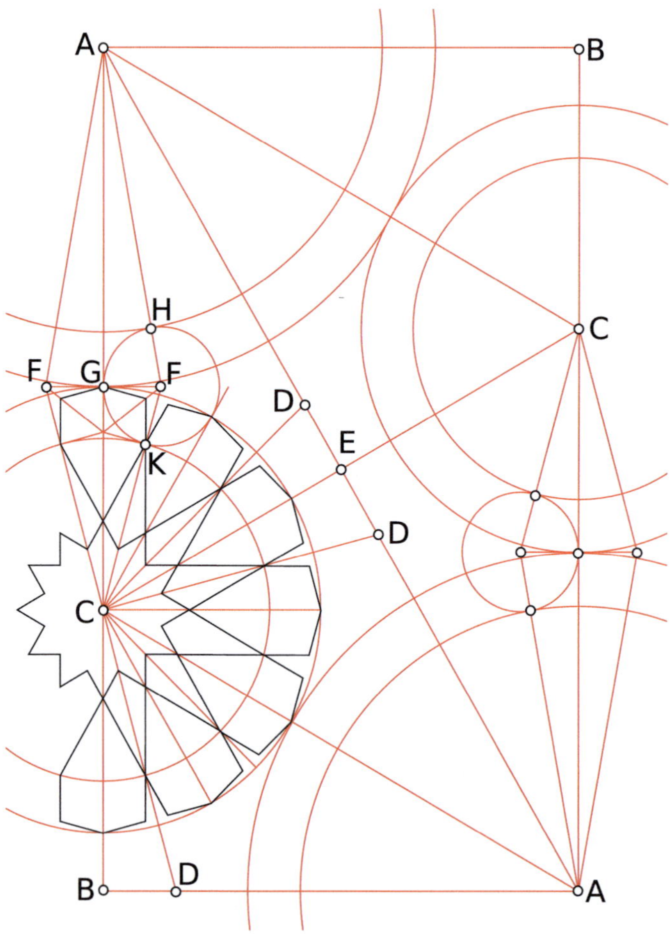

Fig. 16.4 Basic layout

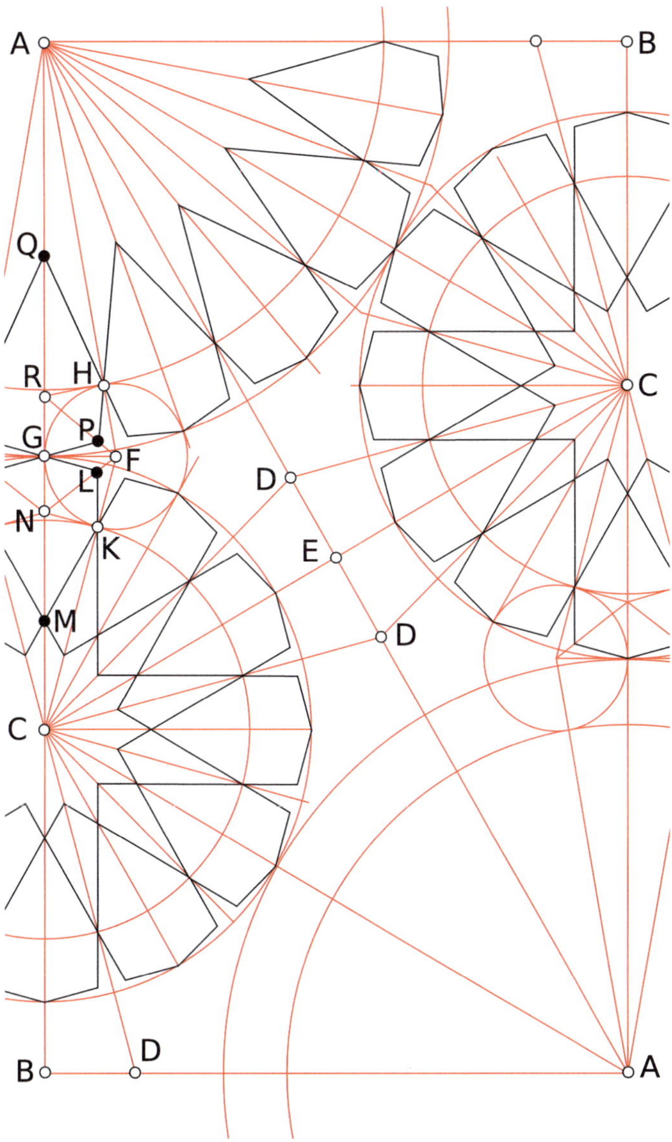

Fig. 16.5 Both rosettes

and G is the middle point between them. At F draw a small circle with radius FG. This circle determines a point H on radius AF. Draw a circle from A through H. The small circle also determines a point K on radius CF. Draw a circle from C through K. Points D will form the centres of the interstitial heptagons.

The 12-pointed rosette at C will have petals with parallel sides. That is, side LK is parallel to radius CG in the petal which has vertices G, L, K and M, and centre N. For

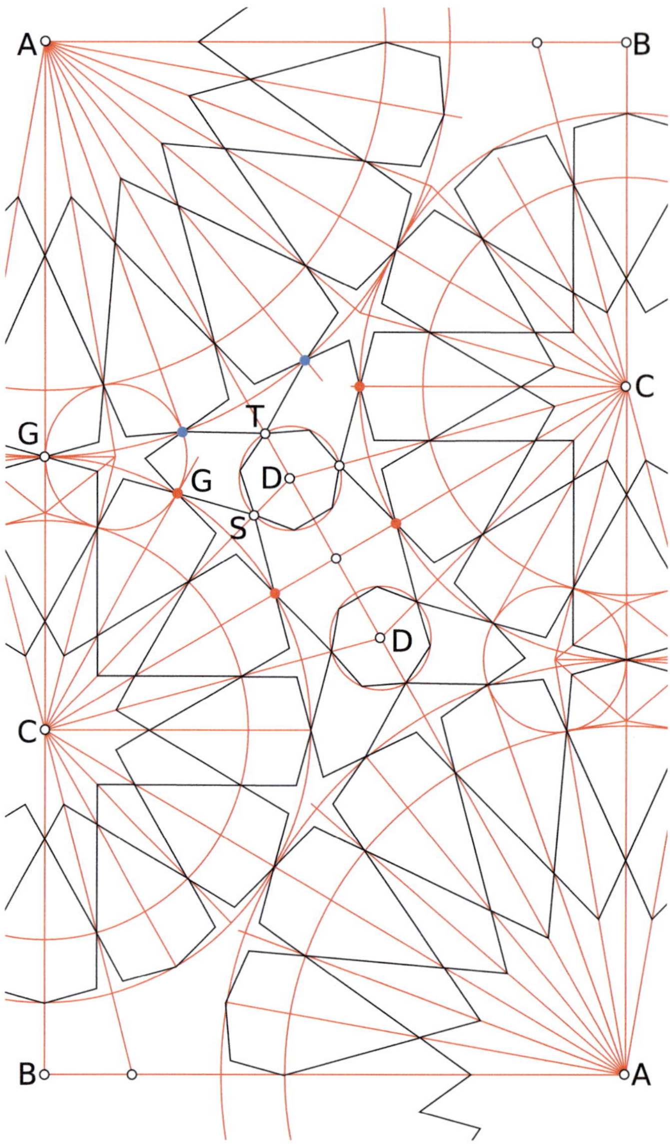

Fig. 16.6 Completion of the interstitial space between 18-pointed and 12-pointed rosettes

other kinds of rosettes, point L (black) can slide along line FM, which is the bisector of angle GFK. Similarly, M can slide along line CN. Construction of the first petal of the 18-pointed rosette. Line GP is continued from the 12-pointed rosette, through G. It intersects the bisector FR of angle GFH at P. Point Q is obtained by making angle AHQ equal to angle FHP. See Fig. 16.5.

With the first heptagon on point D, extend a line G-G until it meets radius CD at point S. Draw a circle on D with radius DS. Draw a *regular* heptagon in this circle, with one of its vertices at T on radius AD. Adjust the nearest vertex to point S so that it coincides with S. We now have a *nearly-regular* heptagon.

Join point S to the nearest rosette vertices marked red, and join point T to the nearest 18-pointed rosette vertices marked blue. Continue lines through adjacent red and blue vertices until they meet, to complete the peripheral 5-pointed star. Lines through blue points are not continuous, nor are those lines through points S and T continuous (Fig. 16.6).

Of course, at this stage, it is possible to make further small adjustments to the heptagon; for instance to minimise the small interlace discontinuity.

16.3 A Comparison

We now have three forms of the pattern: the artefact, the trigonometric version and the manual version. How do they compare?

Firstly, consider the heptagon. The door has inserts which include a heptagon that stands proud of the door. Figure 16.7 includes an exact red pentagon which clearly matches the proud insert (superimposed over a photograph). The green line follows that of the rest of the door pattern, which has angles similar to the regular heptagon

Fig. 16.7 Illustration of the regularity of the heptagon

Fig. 16.8 The completed trigonometric version (Tiling Search Web Site 2017, data168/F142)

but has a much shorter length for the vertical edge on the right. Note that a regular heptagon cannot be drawn with a ruler and compass alone.

Secondly, we consider the vertex angle of the 18-pointed star. For the actual door it is about 42°, for the trigonometric version 44° and for the manual version about

50°. However, this vertex angle can be varied by changing the degree of convergence of the 18-pointed rosette without changing the rest of the pattern.

Thirdly, we consider the relative sizes of the two rosettes. We have noted with the manual method that this ratio is $\tan(10)/\tan(15) = 1.5196$. For the trigonometric version, the radius of the 18-pointed rosette can be calculated from Y in Fig. 16.3. The radius of the 12-pointed rosette is N from the same figure. The ratio is 1.5619. The ratio from the door is about 1.62.

We think that the trigonometric version best encapsulates the design of the door. The final result is shown in Fig. 16.8.

16.4 A Second Mamluk Masterpiece

The metal door just considered has the pattern in a form in which the 'lines' are wide and hence cannot be accurately measured. For the example here, we take a Mamluk minbar in wood which has excellent detail displayed in a high-resolution photograph in Fig. 16.9.

The immediate aspects to note about this pattern is that is has a 9-pointed star with *reflexed petals* and the 12-pointed star with standard petals. The symmetry is $*632$ (*p6m*) which implies that a large area in the photo can be formed by reflection and rotation of a small triangle shown in Fig. 16.10.

One can see from the blue lines that two of the sides of the pentagon subtend an angle of 150°. Hence we make the four sides as shown with the fifth edge of length F subtending an angle of 70°. We are assuming the pentagon is circle-inscribed as with the previous example. Measuring the vertex angle of the 12-pointed star gives a value very close to 72.5°, so we take that value as matching the angle subtended by one of the sides of the pentagon. (For simplicity it is a good idea to make all the angles a multiple of a small amount; 1.25° in this case.)

We can now determine the angles of the pattern. The angles between the $E - E$ edges of the pentagon is 107.5°. Considering the triangle formed by the blue line from the 12-pointed centre to the pentagon and the line GE, the angle at the vertex next to the G is 77.5°. Hence the angle of the middle C's of the petals is 21.25° to the vertical. Hence we can now compute the lengths of A to D by taking A as unity. To aid this, we take X as the length between the two ends of the lines marked C of the petal.

Formula
$A = 1.0$
$B = A * \sin(51.25)/\sin(21.25)$
$X = B * \sin(36.25)/\sin(37.5)$
$C = X/(2 * \cos(31.25))$
$D = X * \sin(52.5)/\sin(53.75)$

Fig. 16.9 Minbar of al-Ghuri mosque, Cairo, dated 1504

Fig. 16.10 Annotated triangle of Minbar

Fig. 16.11 Minbar photograph overlaid with the computer graphic

We can clearly compute the length of the line GED, so G + E is known, but can vary by a small amount. To continue we must assign an estimated value, so take the radius of the inscribed circle for the pentagon, say R. This then gives the lengths E to J. Note that Y is the length of the blue line from the D – D intersection to the centre of the 12-pointed star.

Formula
$R = 1.6$
$E = 2 * R * \sin(36.25)$
$F = 2 * R * \sin(35)$
$Y = D * \sin(111.25)/\sin(15)$
$G = Y * \sin(15)/\sin(38.75) - E$
$H = G * \sin(38.75)/\sin(33.75)$
$I = G * \sin(38.75)/\sin(32.5)$
$J = H * \sin(33.75)/\sin(37.5)$

We still have to determine the size of the kites with edges K and L. One angle of the kites is conventionally 90°, so taking a value of K we have finally:

Formula
$K = 1.2$
$L = K * \tan(53.75)$
$M = (K + E + H) * \tan(33.75) - L$
$N = (I + F + J) * \sin(32.5)/\sin(72.5) - J$

We can now compare the computer drawing with the actual minbar, as shown in Fig. 16.11. The result can be seen to be a good fit. Some processing needed to be undertaken with the photo due to the distortion that arises from a single viewing position. A more precise comparison would require a laser scan of the minbar.

A characteristic of the computer version is that the edge F is about 3% shorter than edge E. This does not seem to be the case with the minbar which therefore leaves the way the minbar was constructed in doubt. Also, the angle of the kite which is drawn as 90° seems slightly larger in the minbar. Photos also reveal that some cross-overs are not quite straight as they surely should be.

Chapter 17
Conclusions

The aim of Part II of this book is to show how to construct Islamic geometric patterns in the optimal way using mathematical techniques. To survey the methods used, we also need to consider approximate methods (which may have been used throughout the past).

There is a school of thought that some mystical, philosophical, or even cosmological process is involved in Islamic design; we see no evidence to support such a claim. On the other hand, the design sometimes does involve a 'Eureka' moment when all the tiles fit together in an elegant fashion. When the patterns such as Fig. 12.1a or 14.5c were first discovered, one can imagine the delight; but the answer was in the geometry.

17.1 Approximate Methods

Jean-Marc Castéra (1999, pp. 98, 99) shows the methods he used to draw the patterns introduced in Chap. 9; he uses a square grid of unit length which can produce good results. However, the vertices of the tiles are not all on the vertices of the grid which implies care is needed in drawing the patterns. It appears this method is used by the currently active Moroccan community. In contrast, in Chap. 9 we take the edge length of the khatem as unity, which would make the edge length of Castéra's grid equal to $1 + \sqrt{2}$. Note also that Paccard appears to show a Moroccan craftsman with a copy of Bourgoin (Paccard 1980, vol. 1, p. 135). Paccard says:

> This carefully bound volume of tracings belongs to maallem Mulay Hafid. The maallems jealously keep these documents which they leave to their successor.

Jules Bourgoin used ruler and compass methods (see next section), but this was augmented with 'adjustments' when problems arose.

Fernández-Puertas (1999, p. 335) considers an alternative construction method for the patterns in Chap. 9; this was apparently used in Islamic art until the Nasrid

B. Wichmann and D. Wade, *Islamic Design: A Mathematical Approach*, Mathematics and the Built Environment 2,
https://doi.org/10.1007/978-3-319-69977-6_17

period. In this method the underlying square of a khatem is divided into 5 parts and the alternate vertices extend by one unit. This forces an error of about 1% compared to the methods used in this book. Such a small error would not be detectable by measuring actual artefacts.

Many patterns can be drawn exactly on a triangular grid, such as Figs. 14.1a and 14.4a. Fernández-Puertas (1999, p. 240) shows how the eastern Umayyad artists divided the edge length of the triangular grid into five parts. This is equivalent to using 1.8 as an approximation to $\sqrt{3}$, which has an error of about 4%.

It appears that Islamic mathematician and astronomer, Abu'l-Wafa Al-Buzjani (940–998) knew of an approximate construction of the regular pentagon which was later copied by Albrecht Dürer. This produces an approximation in which the edge lengths are correct, but the angles are very slightly incorrect; again, this difference could not be detected from measurement of an artefact.

One problem with such approximate methods is the possible accumulation of errors when constructing a large pattern. In the case of a repeating pattern, the rectangle which repeats itself can be used to avoid such. The method described by Castéra avoids the errors that could otherwise arise. Even on a computer, the accumulation can arise; around ten digits of precision are needed in practice.

17.2 Ruler and Compass

Eric Broug (2008) demonstrates the elaborate use of ruler and compass. Unfortunately, there are three problems with this approach. Firstly, the number of construction lines can be confusing for the more complex patterns. Secondly, we know that certain patterns cannot be constructed with a ruler and compass alone (Jagy 2017).[1]

In Fig. 17.1 we see one of the most common of all Islamic patterns. Hence there can be no question as to the proportions of the petals used here; all the classical examples use standard petals. Several instances are known of drawings of this pattern with non-standard petals, including Bourgoin.

The fact that certain polygons and star-polygons cannot be constructed with ruler and compass is barely relevant. As we have noted above, an error of 1% is not detectible in practice and would easily be managed with use of a protractor or similar device.

[1]For a list of tiles that cannot be so constructed, see http://www.tilingsearch.org/sim/sim14.htm.

Fig. 17.1 Bourgoin, Plate 48. The Tomb of Jalal al-Din Hussein, Central Asia, 1152, see (Broug 2008, p 76). Also sites in Turkey, Iran, Iraq, Syria, India and Egypt

17.3 Use of Mathematics

All the mathematics used in the constructions in this book were known to Islamic scholars of the 9th century (see p. 6). Some more obscure mathematical aspects not directly of concern to those producing patterns were not known, such as the details of the seventeen planar patterns (see International Tables for Crystallography 2006 and Appendix A) and also that a few regular polygons cannot be constructed with ruler and compass (see above). Roughly 4% of Islamic patterns cannot be drawn with ruler and compass alone (Tiling Search Web Site 2017, Statistics page).

Hence the question arises as to the extent that mathematics was used by the designers of Islamic patterns. Unfortunately, there is little evidence to come to any conclusions on this issue. The Topkapı Scroll (Necipoğlu 1995) shows some construction lines which clearly indicate the expected understanding of the basic mathematics at that time.

However, many patterns, such as those considered in Chap. 16, require considerable analysis or adjustments to produce an acceptable result. It is only reasonable to assume such adjustments were undertaken manually using visual inspection to obtain an aesthetically pleasing pattern. Nevertheless, it is possibly the case that Islamic scholars were occasionally involved in providing the mathematical analysis that forms the basis of these patterns.

The work in Part II can be seen as an attempt to continue this early analysis augmented with computer based drawing.

In Chaps. 8 and 9, the patterns are purely mathematical and any divergence from the patterns presented here would be an error. In Chap. 10, adjustments are made between the main star and the background; as the number of points of the main star increases, so does the difficulty of making the necessary adjustments. For a 48-

pointed star, the Moroccan craftsmen use many adjustments which would be hard to describe in mathematical terms and no attempt is made to do that here. Unlike the adjustments shown in Sect. 10.2, the background must be changed also in order to give an acceptable result (see Castéra 1999, P. 204).

There are a number of purely mathematical decagonal patterns presented in Chap. 12. However, an important addition to the number could be made by changing the size of the diamonds while retaining the same overall design. This is a general point; many Islamic designs allow one or more parameters to change but too many such designs exist to be presented here.

Many of the 6-fold patterns in Chap. 14 have some adjustable parameters and hence are not completely determined mathematically. A pattern which is determined is Fig. 14.5c.

Chapters 15 and 16 illustrate the key use of mathematics to ensure the framework of each pattern is correct, even if some adjustments are needed.

Finally, it should be noted that many Islamic geometric patterns could not usefully be presented in a purely mathematical form since the number of adjustments to obtain an acceptable result would be excessive. My personal best guess is that somewhere around 5% of patterns are like this.

17.4 The Geometric Time-Line

As indicated earlier, here we are exclusively concerned with the geometric aspects of the Islamic decorative canon. Over time these geometric forms became ever more complex and regional variations developed. This is illustrated by the following example; the khatem (Fig. 4.5) that is found in a Roman floor mosaic later appears in a Samarran house documented (Herzfeld 1923, Fig. 234) as dating from the 9th century. The next development in pattern complexity came with the placing of petals around a star, which features prominently in this book. To our knowledge, the earliest example of this configuration appears in the David collection (Blair and Bloom 2006) and the photograph in Fig. 17.2; this is dated circa 1100.[2] It is worth noting in this case that the artist/craftsman involved could have used a khatem for enclosing the central area, but opted instead for an octagon, which does not actually join the other edges.

The accurate dating of patterns is not always possible. Inscriptions on minbars and buildings sometimes give a date, but otherwise it is necessary to assume the date as that of the building itself. This assumption cannot be made for some mosques, such as the Great Mosque in Damascus which has a long history of changes and additions. Dating on many museum artefacts is made on the basis of style, such as the Mamluk Qur'ān shown in Fig. 1.1.

Taking into account the reservations above, data has been collected of over 750 Islamic patterns from the web site http://tilingsearch.org. A histogram of the data is shown in Fig. 17.3.

---

[2]An example of a rosette about 12 years earlier is provided in (Bonner 2017, p. 43, photo 25.)

Fig. 17.2 Side of a cenotaph, carved wood

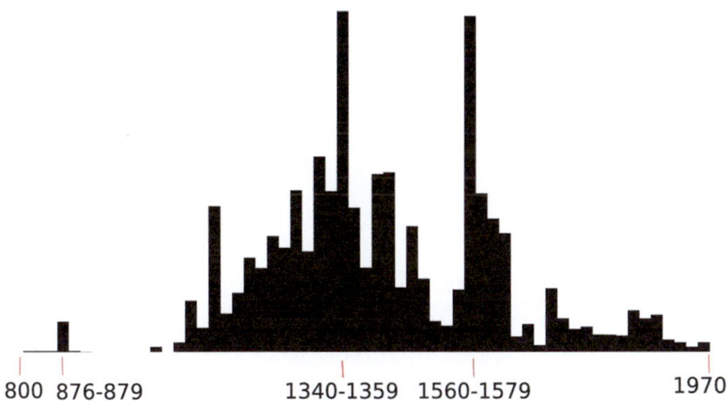

Fig. 17.3 Histogram of pattern usage

The data shown covers the dates from 800 to 1970 and is as follows:

- For each distinct pattern, the earliest known usage contributes to the histogram
- The group of patterns around the period 876–879 are from the Mosque of Ibn Tulun in Cairo. These patterns are part of the original decoration and are in stucco under the arches of the building.
- The group of patterns around the period 1340–1359 are mainly from the Alhambra. Many new patterns are to be found there
- The group of patterns around the period 1569–1579 are mainly from Fatehpur Sikri. Again, many new patterns are to be found there
- This histogram is based on current evidence.

Bearing in mind that the start of the period shown was when the geometry of Euclid was available to the Islamic world (see p. 22), it is surprising that some of the basic patterns arose so late. Even if knowledge of the geometry of Euclid was not used, the development from the classical style was late within the Islamic world. There are two possible reasons for this:

- The actual development of Islamic style could have been a slow process which has certainly happened in other areas of art (not necessarily Islamic)
- Was the wide availability of paper a factor? The use of paper at a later date was clearly important as can be seen from the Topkapı scroll. Moreover, some of the patterns on the scroll do not seem to be present on artefacts which surely demonstrates the use of paper to design patterns (see Fig. 15.5). Necipoğlu states (Necipoğlu 1995, p. 5): it was not, however, until the Mongols arrived in the 1220s that an extensive paper industry developed in Tabriz… This timetable does not seem to be supported in Bloom (2001), Chap. 2, so it is not possible to be sure paper was a key factor in the development of Islamic patterns. In any case, another factor must be the tumultuous events recorded in Chap. 1.

17.5 Pattern Type by Region

In this book we have presented various style of pattern: octagonal in Chaps. 9 and 10; decagonal in Chap. 12; 6-fold in Chap. 14; finally, in Chap. 16 we give an example of an erratic pattern which does not fit into the other styles.

It is well-known that the usage of the various styles varies from region to region. We consider this variation by a journey starting in Spain and going through Morocco, Egypt, Syria, Turkey, Iran, Central Asia to India; left to right in the diagram in Fig. 17.4 giving the percentage of each type. The classification of patterns by type is at least partly subjective.

The dotted lines underneath the main pattern types are rather different and very precise. The red line denotes those patterns having rosettes as defined in Sect. 7.1.1. The other case is when the star has six points and a vertex angle of 60°, since the

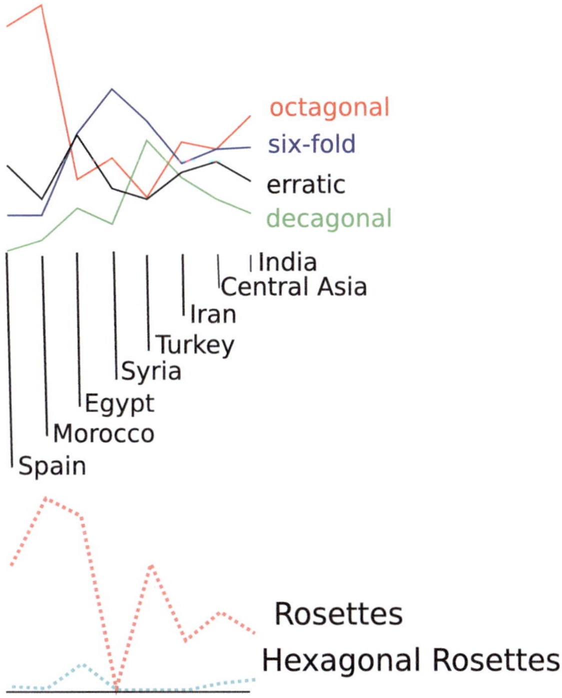

Fig. 17.4 Pattern distribution by country and type

petals can then be hexagons. The numbers of such patterns are shown in the blue line. These could well be Roman or Byzantine.

Note that Morocco has the highest proportion of the octagonal patterns which is hardly surprising in view of the special style of pattern introduced in Chap. 9. The octagonal style is also significant in India, but the details of the style are different. Morocco has almost no decagonal patterns while they are important in Turkey. The erratic patterns are important in Egypt where many are complex as considered in Chap. 16. Syria has a high percentage of 6-fold patterns, which may be partly due to the small total number; the percentages shown are therefore unreliable.

The bottom red line showing rosettes gives the percentage of the total as in the main coloured lines above. Note that the hexagonal rosettes shown with the blue line are also a feature of Roman and Byzantine ornament, which appear most strongly in Egypt.

17.6 A Summary of Methods

Given that here modern tools are being used to construct Islamic patterns, we believe that a mathematical analysis is vital. Two different cases arise according to the results of the mathematical analysis.

17.6.1 Complete Analysis

In Chap. 8 we gave an example where the analysis showed that the pattern was completely determined (although the size of the main star's petals was not easy to determine). In such cases, any divergence from the mathematical solution is an error. Most of the patterns in the octagonal set are also fully determined, see Chap. 9. Such patterns can be quite complex, as that in Fig. 12.13, but must still not diverge from the mathematical solution.

Note that it is not always straightforward to undertake an analysis if some aspect of the pattern is in doubt. For instance, measuring an angle or length (from a photo or an artefact) will not give a mathematically precise value needed for an analysis. For an example of an exact analysis needing some experimentation with the angles, see Sect. 14.5.

17.6.2 Incomplete Analysis

When an analysis is incomplete, i.e. not all angles and lengths are determined, further investigations are needed. In the paragraphs below, we consider some cases which have arisen already with patterns in this book.

With the pattern in Fig. 14.5a, it is clear that the size of the triangle composed of three diamonds and a red polygon can be varied in size. The original craftsman would no doubt have chosen a size which seemed to produce an elegant result. For a modern copy of this design, measurement of the original artefact would suffice.

For another example, consider the Fig. 9.9b. It is clear that the size of the small black square could be varied with corresponding changes to the white and brown tiles. However, there is a strong convention that the edge length of the square is $\sqrt{2}$ times the edge length of the khatem; this is an aspect of the octagonal set (see Chap. 9).

In some cases, the mathematical analysis can be quite complex as with the Mamluk masterpieces in Chap. 16. In the first example the relative size of the rosettes is determined but the interstitial region remains. We have 12-pointed and 18-pointed stars with a heptagon which should appear as regular as possible. Since the numbers 7, 12 and 18 are co-prime, one must expect difficulties; the angles cannot all be ideal in this case, in fact the heptagon is not regular. We have illustrated a manual method in Sect. 16.2. This example also illustrates a key method for using circle-inscribed polygons, as with the heptagon here.

Computer-aided design tools can be used to great effect even when a detailed mathematical analysis cannot be undertaken down to the level of an individual tile. This was illustrated in Fig. 15.11 where the constituent modules can be drawn and then copied to produce an eye-catching result. The effort and skill in undertaking this same process with ceramic tiles must have been substantial.

As far as the classical artisan's methods are concerned, it seems clear that manual techniques with adjustments is critical. Nevertheless, the results of their workmanship are spectacular.

Appendix A
The Symmetry of a Tiling

A.1 Introduction

This book uses notation for the symmetry of tiling patterns defined in (Conway et al. 2008) and in (International Tables for Crystallography 2006) which may not be familiar to many readers. The conventional notation is also noted since this used in the key work: Grünbaum and Shephard (1987). The older notation is somewhat *ad hoc*, whereas the new notation, once understood, has a logical structure and meaning.

The notation uses six symbols, combined with numbers, which are as follows:

- * This denotes a mirror symmetry
- • This indicates that all the symmetries fix a point
- • Also fixes a point, but can be combined with an integer giving a gyration, say 4• (*c4*) as with the pattern in Fig. A.23
- × This is called a miracle which is used in Fig. A.15
- ∞ This infinity symbol is used in Fig. A.19
- O This is the symbol used in Fig. A.17.

For a more detailed description and showing what combinations can appear, see (Conway et al. 2008).

We consider the planar symmetry groups since they are the ones used most in Islamic art.

A.2 The Planar Groups

A.2.1 Mirror lines

We first consider *mirror lines*, which should be easy to see. Consider the pattern in Fig. A.1.

B. Wichmann and D. Wade, *Islamic Design: A Mathematical Approach*, Mathematics and the Built Environment 2,
https://doi.org/10.1007/978-3-319-69977-6

The Great Mosque, Tlemcen, Algeria, 12th century; also in manuscript 1206 (Tiling Search Web Site 2017, data9/H327). No colour in originals, colour added here

Fig. A.1 Example of mirror lines, *632 (*p6m*)

The mirror lines go from one corner of the 6-pointed star to the opposite corner (and extending indefinitely in both directions).

You can see that these mirror lines divide the patterns into triangles, with angles of 30°, 60° and 90°.

It is easy to see that these small triangles have no internal symmetry and hence the mirror lines captures the symmetry of the pattern. Around the vertices of this triangle, we have 6-fold, 3-fold and 2-fold symmetry. The notation (signature) of this symmetry group is *632 (*p6m*). The star indicates the use of mirror lines and the numbers the lines meeting at a point.

We now apply the same logic to the pattern in Fig. A.2.

One set of mirror lines are vertical and horizontal cutting the central 12-pointed star in two. Another set are the diagonals through the central khatem and the other khatem. The triangles formed by the mirror lines have vertices at the centre of main khatem, the center of the other khatem and the centre of the 'khatem' with flat sides. Around the vertices of this triangle, we have 4-fold, 4-fold and 2-fold symmetry. Hence the signature for the symmetry of this pattern is *442 (*p4m*).

Our third example with mirror lines is Fig. A.3.

The pattern contains a 6-pointed star divided into three parts. The mirror lines are the dividing lines of this star. Hence the pattern is divided into triangles with vertices at the centre of the divided 6-pointed star and the two other places where the mirror lines cross. We have 3-fold symmetry around each of these vertices, hence the signature for the symmetry of this pattern is *333 (*p3m1*).

Our fourth example with mirror lines is in Fig. A.4.

Tile from Copenhagen, attribution unknown (Wade 1976, p 93). Also a modern roundel with the same design

Fig. A.2 Second example of mirror lines, *442 (*p4m*)

Fatehpur Sikri, India, carved stone panel. Dated 1565-1605 (Jali screen on balcony) (Tiling Search Web Site 2017, data185/P85). The lines forming a 9-sided polygon cannot be draw with exact regularity using ruler and compass alone (Jagy 2017)

Fig. A.3 Third example of mirror lines, *333 (*p3m1*)

Golestan Palace in Tehran. original colour (Tiling Search Web Site 2017, data17/P30)

Fig. A.4 Fourth example of mirror lines, *2222 (*pmm*)

Here the mirror lines are vertical and horizontal. One set of vertical lines goes through the white polygons, the other through the 8-pointed star which appears between the four white polygons. The horizontal ones go through either the 8-pointed stars only or through the 8-pointed star and square. Hence the pattern is divided into rectangles with 2-fold symmetry at each vertex. Hence the signature for this pattern is *2222 (*pmm*).

A.2.2 Rotations

Here we consider points in the pattern about which a rotation can be undertaken to leave the pattern unchanged. Corresponding to each of the four patterns above, we have a pattern with the same number of rotations. All these patterns are different from their mirror image (and have no mirror lines).

In Fig. A.5, the centres of the rotations are the center of the hexagons, centres of the 6-pointed star and the point midway between two neighbouring hexagons. Hence this is 632 (*p6*).

In Fig. A.6, the centres of the rotations are the center of the 8-pointed star, centres of the square and the point midway between two neighbouring 8-pointed stars. Hence this is 442 (*p4*).

Itmad ud Daula, Agra, 1626 and also Fatehpur Sikri, Salim Chishti's Tomb, 1581 (Tiling Search Web Site 2017, data8/PG351)

Fig. A.5 Rotations of order 6, 632 (*p6*)

Alhambra, Mexuar Patio corridor. No colour in the original. Has a left-hand and right-hand versions (Tiling Search Web Site 2017, data159/P051)

Fig. A.6 Rotations of order 4, 442 (*p4*)

Alhambra, Patio de los Arrayanes, 1354-1362. Colours vary in the original—a selection used here (Patterns in Islamic Art web site 2017, SPA 0108)

Fig. A.7 Three centres of rotation, 333 (*p3*)

In Fig. A.7, the centres of rotation are at the centre of the 6-pointed star, centre of the white hexagon and where three green polygons meet.

Figure A.8 has four different rotations of order 2: this can be seen from the rotations which preserve the colours of the individual tiles. Hence this is 2222 (*p2*).

A.2.3 Glide Reflections

In Fig. A.9, we have two vertical mirror lines through the middle dark green polygon and red large star. This case is different from the mirror line examples seen previously since the area between the mirror lines is not finite.

We have now considered the symmetries which involve *only* mirrors or *only* rotations. We have eight more cases in which both symmetries are present.

In Fig. A.10, we have mirror lines vertically and horizontally through the octagon. We can also rotate by 180° around the centre of the octagon. A rotation of order 4 is present about the point at which four black polygons meet. The notation shows all these three properties.

Figure A.11 has a rotation of order 3 around the green 6-sided polygon, a mirror line by extending the lines through the 6-pointed star, and a rotation around the centre of the 6-pointed star.

Tomb of Itmad ud Daula, Agra, 1626 (Patterns in Islamic Art web site 2017, IND 0428)

Fig. A.8 Four different rotations of order 2, 2222 (*p2*)

Topkapı scroll, See (Necipoğlu 1995, p. 313). No actual use of this pattern known apart from the scroll

Fig. A.9 Decagonal pattern in strips, ** (*pm*)

Friday Mosque, Yazd, 1325-34 and Ibril, Iraq (minaret) 1190-1232 (Tiling Search Web Site 2017, data19/R14). Colour from Yazd

Fig. A.10 Octagon with square pattern of order four, 4*2 (*p4g*)

Figure A.12 has mirror lines along the major and minor axes of the black 8-sided figure. There is also a rotation of order 2 round the point midway between the two large black stars and a rotation of order 2 at the centre of the black 8-sided polygon.

Figure A.13 has a mirror line vertically in the middle of the large stars. The pattern can be rotated 180 degrees about the middle of the brown 10-sided polygon. Also a similar rotation at the point in the middle of the white 8-sided polygon which are to the left and right of the main star.

In Fig. A.14 we have two rotations by 180°: one about the centre of the white polygon and the other where two white polygons meet. But there is another symmetry: consider a brown polygon. It can be moved along horizontally and then reflected in the line between the brown polygons. This symmetry operation is called a *glide-reflection* and is denoted by ×. Also, the white polygons can be move up to the next row of white polygons and then reflected giving another glide-reflection Hence the signature for the pattern above.

In Fig. A.15, the large mainly yellow decagons all have the inner pentagon in the same orientation. The rows of these clearly alternate. There are two vertical mirror lines through the two rows of decagons. There is also a glide-reflection moving the decagons from one row to the next. The glide-reflections are vertical, alternating with the vertical mirror axes.

The 12-sided polygon which has six 3-pointed stars within it. Mosque of Rustem Pasha, Istanbul 1560 (Denny 1998, Plate 11). Balustrade of minbar, no colour in original, so colour added here

Fig. A.11 Pattern with 12-sided polygon, 3*3 (*p31m*)

This decagonal pattern has a 10-point star and regular pentagons. Earliest known example from Isfahan, 1088 (Seherr-Thoss 1968, Plate 10). Does not seem to occur in Western Islam. This design features in (Bourgoin 1879, Plate 175)

Fig. A.12 Widespread decagonal pattern, 2*22 (*cmm*)

Mosque in the Citadel, Cairo 1284-5, also in a Victorian mosque in Cairo (Tiling Search Web Site 2017, data19/M2)

Fig. A.13 Vertical decagonal pattern, 22* (*pmg*)

No Islamic source known. The pattern is produced with tiles used in Islamic decagonal designs. See (Rigby and Wichman 2006)

Fig. A.14 Pattern with alternating strips, 22× (*pgg*)

Qur'ān from National Library of Cairom Fig,13.7. Mamluk period 1250-1517. (UNESCO Memory of the World)

Fig. A.15 11 pentagons inside a regular decagon, *× (*cm*)

No Islamic source known. Pattern produced from decagonal Islamic tiles. See (Rigby and Wichman 2006)

Fig. A.16 Vertical strip pattern, ×× (*pg*)

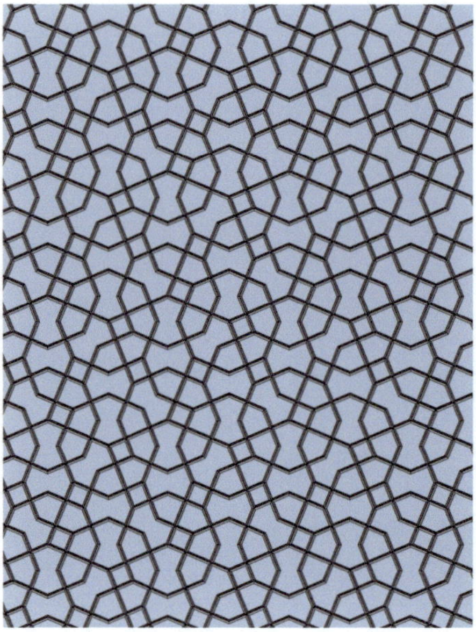

Mausoleum of Muhammad Ghaus, Gwalior, India, 1565. Part of a large jali screen (Broug 2013, p 105, Fig. 4.35)

Fig. A.17 Pattern with overlapping octagons, O (*p1*)

In Fig. A.16 we have no mirror lines. However, the brown polygon can be moved both up and down and flipped left-to-right. In other words, we have two glide-reflections. Hence the signature for this pattern.

In Fig. A.17 we have no mirror lines, rotations or glide-reflections. This gives the special symbol as shown.

A.3 The Frieze Groups

The symmetry groups of frieze patterns are introduced and illustrated in Figs. A.18 and A.19.

Fig. A.18 First Frieze pattern, *22∞ (*pma2*)

Ulu Mosque, Eski Malatya, Turkey, 1224. This uses decagonal tiles with very wide interlacing, original colours (Broug 2013, p 64, Fig. 2.65)

Fig. A.19 Second Frieze pattern, 2*∞ (*pmm2*)

Karatay Madrasah, Konya, Turkey. This pattern also uses decagonal tiles with wide interlacing. Bernard O'Kane states: Formerly a school for religious students, this is now a museum of ceramics. Built by the amir Jalal al-Din Karatay in 1251

A.4 The Circular Groups

Known Islamic examples appear in Figs. A.20–A.32.

Fig. A.20 2-fold symmetry, *2• (*d2*)

Alhambra Museum, Wooden screen (Arte islámico en Granada 1995)

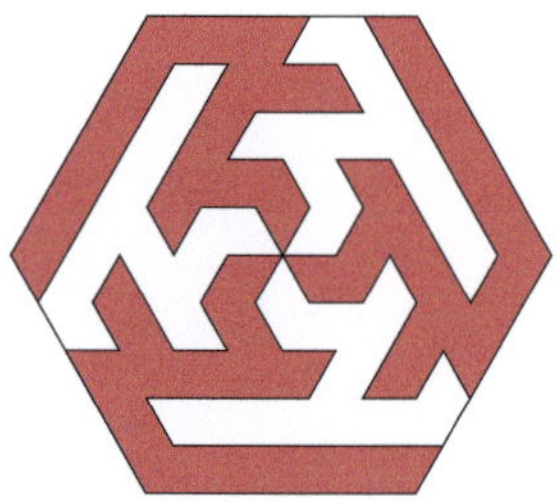

Fig. A.21 3-fold symmetry, *3• (*c3*)

Illustration in the Topkapı scroll (Tiling Search Web Site 2017, data201/ Sakkal 2010)

Fig. A.22 4-fold symmetry, *4• (*d4*)

Sayyed mosque, Isfahan, Iran, 1840 (Patterns in Islamic Art web site 2017, IRA 1617)

Fig. A.23 4-fold symmetry, 4• (*c4*)

Amiriya Madrasa, The Yemen (Wichmann and Rigby 2009). This pattern spells out 'Ali' in square Kufic script

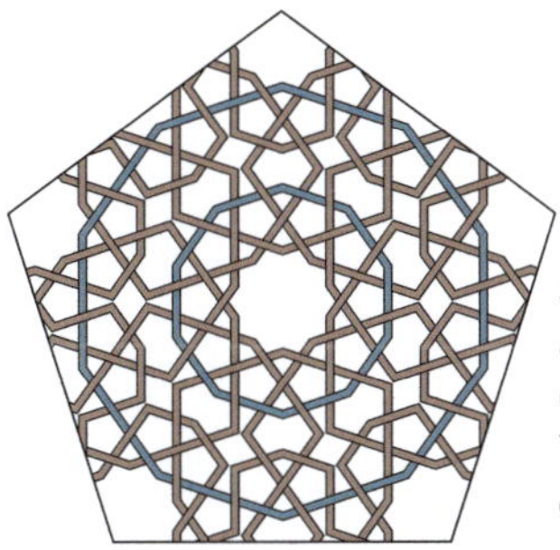

Fig. A.24 5-fold symmetry, *5• (*d5*)

Sifahidiye Madrasah, Sivas, Turkey, 1277 (Tiling Search Web Site 2017, data188/SIVAS3)

Fig. A.25 5-fold symmetry, *5• (*d5*)

Ali Qoli Aqa mosque, Isfahan, Iran. 1122 (Maher al-Naqsh 1983, vol. 4, p. 155)

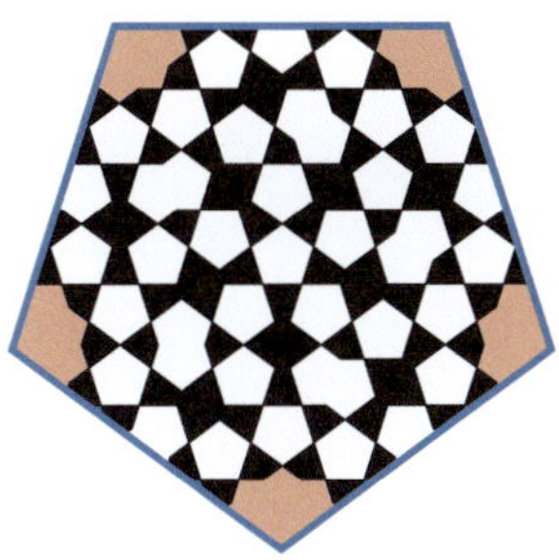

Fig. A.26 5-fold symmetry, 5• (*c5*)

Imamzeda Darbi Islam, Isfahan, Iran, 1453 (Patterns in Islamic Art web site 2017, IRA 0907)

Fig. A.27 6-fold symmetry, *6• (*d6*)

Sircali Madrasah, Konya, Turkey, 1242 (Tiling Search Web Site 2017, data194/KONYA1)

Fig. A.28 6-fold symmetry, 6• (*c6*)

Janpanah, Abbasi Courtyard, Mashhad, Iran (Maher al-Naqsh 1983, vol. 1, p. 116)

Photograph of Iranian roundel (Maher al-Naqsh 1983, vol. 4, p. 149)

Fig. A.29 7-fold symmetry, *7• (*d7*)

Nejjarine Palace, Fez, Morocco (Tiling Search Web Site 2017, data196/N4)

Fig. A.30 8-fold symmetry, *8• (*d8*)

Fig. A.31 10-fold symmetry, *10• (*d10*)

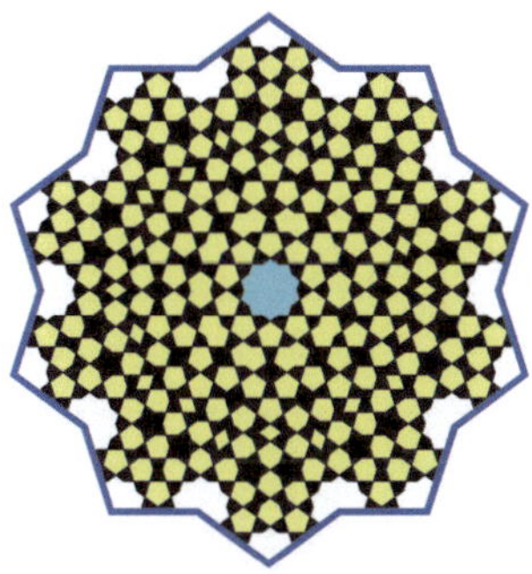

Charbagh Madrasah, Isfahan, Iran. 1122 (Maher al-Naqsh 1983, vol. 1, p. 158)

Fig. A.32 12-fold symmetry, *12• (*d12*)

Caravanserai of Amir Qawsun, Cairo (Sutton 2007, p. 183)

A.5 Notations Compared, with Frequencies

The table below gives a list of the symmetry groups used in Islamic geometric patterns (Sects. A.2, A.3, A.4). The frequency is given in number per thousand (for each distinct pattern type). Of course, the Islamic artists would not have known all the seventeen planar symmetry groups.

Symmetry	*Frequency*	*Figure*
*× (*cm*)	1	A.15
2*22 (*cmm*)	112	A.12
O (*p1*)	1	A.17
2222 (*p2*)	2	A.8
333 (*p3*)	1	A.7
3*3 (*p31m*)	9	A.11
*333 (*p3m1*)	1	A.3
442 (*p4*)	35	A.6
4*2 (*p4g*)	65	A.10
*442 (*p4m*)	446	A.2
632 (*p6*)	26	A.5
*632 (*p6m*)	198	A.1
×× (*pg*)	0	A.16
22× (*pgg*)	0	A.14
** (*pm*)	5	A.9
22* (*pmg*)	1	A.13
*2222 (*pmm*)	54	A.4
∞∞ (*p111*)	0	
∞× (*p1a1*)	0	
∞* (*p1m1*)	0	
*∞∞ (*pm11*)	0	
22∞ (*p112*)	0	
*22∞ (*pma2*)	1	A.18
2*∞ (*pmm2*)	1	A.19
*2• (*d2*)	1	A.20
*3• (*c3*)	1	A.21
*4• (*d4*)	2	A.22
4• (*c4*)	1	A.23
*5• (*d5*)	2	A.24
5• (*c5*)	1	A.26
6• (*c6*)	8	A.28
*6• (*d6*)	1	A.27
*7• (*d7*)	1	A.29
*8• (*d8*)	7	A.30
*10• (*d10*)	5	A.31
*12• (*d12*)	1	A.32

The circular groups are only listed for those which occur in known Islamic patterns (Fig. A.24).

Appendix B
Key Dates

Date	*Event*
410	Fall of Rome
570	Birth of Muhammad
622	Muhammad's hegira with his followers from Mecca to Medina
632	Death of Muhammad
632–4	Caliph Abu Bakr quells uprisings in Arabia, begins invasions of Mesopotamia and Palestine
635	Damascus surrenders to the Muslims
639	Arab capture of Syria and Egypt
653	Compilation of the revelations of the Prophet Muhammad and official edition of Qur'an
661–680	Caliph Muawiya I begins second great period of expansion
691	Construction of the Dome of the Rock in Jerusalem
705–15	Caliph Walid I; Conquest and Islamicisation of Central Asiatic centres of Bukhara and Samarkand
711–4	Arab Muslim forces overrun Spain
712	Arab Muslim forces cross the Jaxartes and advance to Kashgar
713	Arab Muslim forces invade Indus valley and take Multan
732	Frankish and Burgundian forces defeat the Arabs in France
749	Establishment of Abbassid Dynasty
755–88	Abd ar-Rahman I founds Spanish Umayyad dynasty
762	The founding of Baghdad by the Caliph al-Mansur
785	Construction of Umayyad Mosque in Cordova
786	Caliph Harun ar-Rashid; high point of Abbasid Caliphate in Baghdad
810	Golden Age of Arabic Science begins; translation of Euclid into Arabic
827	Caliph al-Mamun establishes Mutazilism as a state orthodoxy (which encouraged speculative dogma in Islam)

B. Wichmann and D. Wade, *Islamic Design: A Mathematical Approach*, Mathematics and the Built Environment 2,
https://doi.org/10.1007/978-3-319-69977-6

827	Arab conquest of Sicily
833	Transference of seat of Caliphate to Samarra; Caliph al-Mutasim becomes puppet of his Turkish guards
873	Death of Hunain ibn Ishaq (who was responsible for the greatest era of translation from Classical Greek texts)
876–9	Ibn Tulun mosque in Cairo built
912–61	Abd ar-Rahman II: height of Spanish Umayyad rule (adopts title of Caliph 929)
956	Seljuk Turks embrace Islam
996–1030	Mahmud of Ghazna conquers North West India
1031	Collapse of Umayyad rule in Spain and disintegration into petty states (Taifa)
1063–72	Turkish forces, under Seljuk Sultan Alp Arslan, break into eastern Asia Minor
1096–9	First Crusade; Jerusalem conquered, becomes Latin Kingdom
1100	First Islamic pattern with petals
1219–24	Mongols, under Genghis Khan, ravage Transoxiana and Khurasan
1236	Great Mosque in Cordova converted to a church
1250	Rule of the Mamluks (Turkish military slaves) begins in Egypt
1258	Fall of Baghdad to the Mongols (under Hulagu); end of the Abbasid Caliphate
1281	Beginning of Ottoman rule in Anatolia
1398	Mongols, under Timur, attack Northern India
1314	Building of the Alhambra Palace started
1453	Constantinople taken by the Ottomans
1492	The last remaining Muslim enclave in Spain, the Nasrids of Granada, falls to Christian forces leading to Muslim expulsion
1501	Safawid rule begins in Iran; capital established at Isfahan
1501–10	Uzbeks conquer Central Asia; principle cities Bukhara and Samarkand
1526	Babur begins Mughal rule in India; capital Agra
1556	The Mughal Emperor Akbar assumes power
1571–3	Building of Fatehpur Sikri
1500–1800	The Age of European expansion into Islamic lands

(Some dates are approximate)

Appendix C
Glossary

- *ashlar*
  Square dressed stone.
- *base pattern*
  The abstract version of the pattern ignoring aspects such as colour and the form of the edges. (Abas and Salman 1993, p. 70) uses uses the term 'ground symmetry' for the same concept.
- *dart*
  A concave 4-sided polygon whose edges are AABB. For an example, see Fig. 9.10 (6).
- *edge-to-edge*
  A property of the tiling pattern in which two neigbouring tiles have just one edge in common. Neither edge continues in the same straight line.
- *Fritware*
  also known as Islamic stone-paste, is a type of pottery in which frit is added to clay to reduce its fusion temperature. As a result, the mixture can be fired at a lower temperature than clay alone.
- *girih*
  See: Umm al-Girih.
- *guilloche*
  repetitive architectural patterns of intersecting or weaving designs. Roman designs look like a rope.
- *hegira*
  is the migration or journey of the Islamic prophet Muhammad and his followers from Mecca to Medina in June 622 CE.
- *interstitial*
  In a pattern having rosettes, the regions that are outside the rosettes themselves.
- *kashi*
  Persian tile work or mosaic

B. Wichmann and D. Wade, *Islamic Design: A Mathematical Approach*, Mathematics and the Built Environment 2,
https://doi.org/10.1007/978-3-319-69977-6

- *khatem*
  This is the name given to the regular star polygon with 8 points and a vertex angle of 90°. For an example, see the yellow polygon in Fig. 9.3a.
- *kite*
  A convex four-sided polygon whose edges are AABB. For an example, see Fig. 9.10(1).
- *madrasah*
  Islamic school. This spelling is used to avoid local variations.
- *Maghreb*
  The area sometimes referred to as 'Western Islam', being Moorish Spain, Morocco, and North Africa west of Egypt.
- *mihrab*
  semicircular niche in the wall of a mosque that indicates the qibla; that is, the direction of the Ka'ba in Mecca and hence the direction that Muslims should face when praying.
- *minbar*
  Pulpit in a mosque, often made of wood and decorated
- *Moroccan style*
  We give this name to patterns which mainly contain octagonal shapes, as noted in Fig. 9.10.
- *Mudéjar*
  Islamic style produced in Spain when controlled by Christian kings.
- *petal*
  Part of a rosette; the 6-sided polygons next to the kites or the star polygon.

  standard petal
  petal whose four sides furthest from the star-polygon are of the same length.

- *roundel*
  A motif which has circular symmetry as in the three examples in Fig. 12.2.
- *sgraffito*
  technique either of wall decor, produced by applying layers of plaster tinted in contrasting colours to a moistened surface, or in ceramics, by applying to an unfired ceramic body two successive layers of contrasting slip, and then in either case scratching so as to produce an outline drawing.
- *wakala*
  urban warehouse with dwellings.
- *Umm al-Girih*
  Persian for mother of pearl, but denotes a method of drawing decagonal patterns, see page 157.
- *zellij*
  Mosaic or tilework made from individually chiseled geometric tiles set into a plaster base.

Appendix D
Copyright

The list is alphabetical under the copyright holder.

Copyright holder	*Figure*
Mahmood Maher al-Naqsh	Fig. A.29
ACR Edition	Fig. 11.3c
Bakanov, Alexandr /stock.adobe.com	Fig. at the start of Chapter 3
Crossling, Nick	Fig. 5.5, 8.1, 8.5, 13.2a
bpk / Museum für Islamische Kunst, SMB / Johannes Kramer,	Fig. 3.1
David Collection, Copenhagen, 1/2004, photographer: Pernille Klemp	Fig. 17.2
Henry, Richard	Fig. 15.9a
Karssenberg, Goossen	Fig. 13.4
Metropolitan Museum of Art	Figs. 5.7, 15.2a,
McMorrow, Brian	Fig. 5.1
Museum of the History of Science, University of Oxford, item 45307	Fig. 2.1
National Library and Archives of Egypt	Fig. 1.1, Fig. 13.7a
Soliman, Ayman	Fig. 11.3b, Fig. 11.7a
	Notice on page 215, Figs. 16.1, 16.2
Trustees of the Chester Beatty Library, Dublin	Fig. 11.3a
Wade, David	All remaining photos
Wichmann, Brian	All remaining graphics

B. Wichmann and D. Wade, *Islamic Design: A Mathematical Approach*, Mathematics and the Built Environment 2,
https://doi.org/10.1007/978-3-319-69977-6

References

Abas, S.J., and A.S. Salman. 1993. *Symmetries of Islamic Geometrical Patterns*. Singapore: World Scientific.

Abd-al-Wahhab, Hasan. 1946. *Tarikh al-masajid al-athariyah*. 2 vols. Cairo. (In Arabic.)

Alamy. A Web Site For Photographs. http://alamy.com.

Al-Khalili, Jim. 2012. *Pathfinders: The Golden Age Of Arabic Science*. Allen Lane.

Arte islámico en Granada. 1995. Arte islámico en Granada: Propuesta para un museo de la Alhambra. Exhibition 1 April – 30 September 1995, Granada: Palacio de Carlos V, La Alhambra. Exhibition catalogue. ISBN 9788481511352

Aslanapa, Oktay. 1971. *Turkish art and architecture*. London: Faber. ISBN 0571087817.

Balmelle, Catherine. 2002. *Le décor géométrique de la mosaique romaine*. Paris: Picard.

Blair, Sheila S. and Jonathan M. Bloom. 2006. *Cosmophilia: Islamic Art From The David Collection, Copenhagen*. Boston.

Blair Sheila, S., and Bloom Jonathan M. 2009. *Grove Encyclopedia of Islamic Art & Architecture*. USA: OUP.

Bloom, Jonathan M. 2001. *Paper before Print: The History and Impact of Paper in the Islamic World*. London: Yale University Press.

Bonner, Jay. 2017. *Islamic Geometric Patterns: Their Historical Development and Traditional Methods of Construction*. Springer.

Bourgoin, Jules. 1879. *Les Eléments de l'art arabe: le trait des entrelacs*. Paris. (Patterns available as a Dover reprint,)

Broug, Eric. 2008. *Islamic Geometric Patterns*. Thames & Hudson.

Broug, Eric. 2013. *Islamic Geometric Design*. Thames and Hudson.

Castéra, Jean-Marc. 1999. *Arabesques: Decorative Art in Morocco*. Paris: ACR Edition.

Conway, J.H., H. Burgiel and C. Goodman-Strauss. 2008. *The symmetries of things*. A K Peters Ltd.

Creswell, K.A.C. 1952–1959. *The Muslim Architecture of Egypt*. 2 vols. Oxford.

Cromwell, Peter. 2016. Modularity and Hierarchy in Persian Geometric Ornament. *Nexus Network Journal* 18 (1): 7–54.

Degeorge, Gérard and Yves Porter. 2002. *The Art of the Islamic Tile*. Paris.

Denny, Walter. 1998. *Gardens of paradise*. Ertug and Kocabiyik.

Fernández-Puertas, Antonio. 1999. *The Alhambra: Volume I—from the ninth century to Yusuf I (1354)*. SAQI Books.

B. Wichmann and D. Wade, *Islamic Design: A Mathematical Approach*, Mathematics and the Built Environment 2,
https://doi.org/10.1007/978-3-319-69977-6

Golombek, Lisa and Donald Wilber. 1988. *The Timurid architecture of Iran and Turan*. Princeton.
Grünbaum, B., and G.C. Shephard. 1987. *Tilings and Patterns*. New York, NY: W. H. Freeman & Co.
Grünbaum, B., and G.C. Shephard. 1992. Interlace Patterns in Islamic and Moorish Art. *Leonardo* 25 (3/4): 331–339.
Herzfeld, E. 1923. *Die Ausgrabungen von Sammarra I*. Berlin: Der Wandschmuck der Bauten von Samarra und seine Ornamentik.
Hessemer, F M. 1990. *Historic Designs and Patterns in Color*. Dover reprint.
Hill, Derek and Oleg Grabar. 1964. *Islamic architecture and its decoration, A.D. 800–1500: A Photographic Survey*. Chicago: University of Chicago Press.
International Tables for Crystallography. 2006. Vol. A, Space-group symmetry, 1st online edition, ed. by Th. Hahn. Chester: International Union of Crystallography. https://doi.org/10.1107/97809553602060000100.
Jagy, Will. 2017. *What Islamic tiling patterns are constructible?* http://mathoverflow.net/questions/35526.
Jones, Owen and Jules Goury. 1837. *Plans, elevations, sections, and details of the Alhambra.....* London.
Jones, Owen. 1856. *The Grammar of Ornament*. Dover. (reprint) Original version available on the Internet.
Khaldun, Ibn. 1967. *The Muqaddimah: An Introduction to History*, ed. N.A. Dawood; trans. Franz Rosenthal. Princeton.
Kühnel, Ernst. 1977. *Arabesques*. Verlag für Sammler.
Lee, Tony and Ayman Soliman. 2014. *The Geometric Rosette : Analysis of an Islamic Decorative Motif*. http://www.tilingsearch.org.
Lee, A.J. 1987. Islamic star patterns. *Muqarnas* 4: 182–197.
Locher, J.L. (ed.). 1992. *Escher — The Complete Graphic Work*. Thames and Hudson.
Lu, P.J., and P.J. Steinhardt. 2007. *Decagonal and quasi-crystalline tilings in medieval Islamic architecture*. *Science* Feb 23 2017; 315 (5815):1106–10.
Maher al-Naqsh, Mahmood. 1983. *Design and execution of drawing in Iranian tilework*. Reza Abbasi Museum: Islamic period. (In Persian).
Moustafa, A., and S. Sperl. 2014. *The Cosmic Script, Sacred Geometry and the Science of Arabic Penmanship*. London: Thames & Hudson.
Neal, D.S. 1981. *Roman mosaics in Britain*. Society for the Promotion of Roman studies.
Necipoğlu, Gülru. 1995. *The Topkapı scroll: geometry and ornament in Islamic architecture*. The Getty Center for the History of Art and the Humanities.
O'Kane, Bernard. 2016. *The Mosques of Egypt*. Cairo: AUC.
O'Leary, De Lacy. 1949. *How Greek Science Passed to the Arabs*. Routledge and K. Paul.
Paccard, Andrè. 1980. *Traditional Islamic craft in Moroccan architecture*. France: Saint-Jorioz.
Pajares-Ayuela, Paloma. 2002. *Cosmatesque ornament: flat polychrome geometric patterns*. London: Thames and Hudson.
Patterns in Islamic Art. 2017. The web site: www.patterninislamicart.com. Specific references (e.g., EGY 0512) provide terms to be inserted into the web site search box.
Pope, Arthur Upham. 1965. *Persian Architecture*. London: Thames and Hudson.
Rigby, John and Brian Wichmann. 2006. Some patterns using specific tiles. *Visual Mathematics* 8(2).
Sakkal, Mamoun. 2010. *Square Kufic Calligraphy in Modern Art, Transmission and Transformation*. Washington: University of Washington.
Saunders, J.J. 1965. *A History of Medieval Islam*. Routledge and Kegan Paul.
Schattschneider, D. 1990. *M C Escher — Visions of Symmetry*. W H Freeman and Company.
Seherr-Thoss, Sonia, P. 1968. *Design and Color in Islamic Architecture: Afghanistan, Iran Turkey*. USA: Smithsonian Institution.
Shafai, Javad. 1977. *On decoration in Persian architecture and wood carving*. Tehran. (In Persian)
Singer, Lynette. 2008. *The Minbar of Saladin*. London: Thames & Hudson.

Smith, E.W. 1894–1899. *The Moghul Architecture of Fathpur-Sikri*, 4 vols. Archaeological Survey of India. Rpt. Delhi, 1985.

Sutton, Daud. 2007. *Islamic Design — a genius for geometry*. London: Wooden books.

Tabbaa, Yasser. 2001. *The Transformation of Islamic Art during the Sunni Revival*. Washington: University of Washington Press.

Tiling Search Web Site. 2017. www.tilingsearch.org. Data references refer to a specific web site URL, e.g., data162/P007 refers to www.tilingsearch.org/HTML/data162/P077.html.

UNESCO. Memory of the World: The National Library of Egypt's Collection of Mamluk Qur'an Manuscripts. http://www.unesco.org/new/en/communication-and-information/memory-of-theworld/register/full-list-of-registered-heritage/registered-heritage-page-8/the-national-libraryof-egypts-collection-of-mamluk-quran-manuscripts/.

Victoria and Albert Museum collections website. 2017. Individual items are referenced by means of an object number. References to the web site 1178103 to refer to the URL http://collections.vam.ac.uk/item/O1178103.

Wade, David. 1976. *Pattern in Islamic Art*. London: Studio Vista.

Wichmann, B.A. 2001. *The World of Patterns*. Singapore: World Scientific. ISBN 981024619.

Wichmann, Brian and John Rigby. 2009. Yemeni Squares. Leonardo. 42 (2): 156–162.

Index

A
Abas, S., 211, 217
Abbasid, 5, 20, 21, 34, 35
Abd al-Gani al-Fakhri, Cairo, 132
Abd al-Rahman Katkhuda, Cairo, 121
Abd-al-Wahhab, Hasan, 217
Abd-ar-Rahman, 13
Abu Bakr, 3
Aesthetic, 26, 31, 34, 39, 55, 185
Afghanistan, 10, 11, 81, 125
Agra, India, 104, 118, 124, 195, 197
Ahmad b. Tulun, 7
Akbar, 12, 52, 152
Al-Ashraf Barsbay, Cairo, 66
Al-Azhar, 42, 140
Al-Djanna, 30
Al-Fadawiyya, Cairo, 66, 137, 138
Al-Farabi, 20, 21
Al-Faradis, 56
Al-Kahira, 7
Al-Kindi, 20
Al-Ma'mun, 18
Al-Mu'ayyad, Cairo, 88, 108
Al-Mutasim, 9
Al-Otrush, 56
Al-Rifa'i, Cairo, 108
Alcázar, 89
Aleppo, 10, 19, 56, 131
Alexandria, 20
Algeria, 14, 192
Alhambra, 14, 17, 45, 55, 62, 65, 69, 76, 78—81, 105, 106, 110, 112, 146, 195, 196, 204
Ali Qoli Aqa mosque, Isfahan, 118, 122
Alids, 7
Allah, 4, 26, 34
Almohads, 14
Almoravid, 14
Amir Qijmas al-Ishaqi, Cairo, 106
Amir Taz Palace, Cairo, 104
Anatolia, 58
Aqbughawiyya, Cairo, 104
Aquilee, 137
Ar-Razi, 21
Arabesque, 36, 38, 41, 44, 54, 55
Arabia, 3
Arabic, 3, 6, 10, 11, 14, 17, 18, 22, 63
Architecture, 8, 11, 31, 35, 37, 41, 53
Aristotle, 19
Ashlar, 41, 211
Aslanapa, Oktay, 217
Astrolabe, 18
Astronomy, 22
Ayyubid, 11, 36, 41

B
Babur, 12
Baghdad, 5, 9, 10, 17, 21, 34, 35, 38, 47, 54
Balkans, 11
Balmelle, Catherine, 217
Band, 103, 109
Base pattern, 105, 211
Ben Youssef, 100
Berber, 7, 13, 44
Blair, Sheila, 217
Bloom, Jonathan, 39, 217
Book, 31, 41, 47
Border, 88, 105, 108
Border, irregular, 112

B. Wichmann and D. Wade, *Islamic Design: A Mathematical Approach*, Mathematics and the Built Environment 2,
https://doi.org/10.1007/978-3-319-69977-6

Bou Inaniya, Morocco, 77
Bourgoin, Jules, v, 121, 131, 132, 140, 143, 146, 199, 217
Brickwork, 34, 42, 53, 57
Britain, 15
British Library, 142
Broug, Eric, 184, 217
Bukhara, 9, 56, 125, 146
Buyid, 7
Byzantine, 4, 5, 28, 34, 41, 53, 137, 189

C
Cairo, 7, 10, 19, 32, 42, 46, 52, 65, 66, 103, 117, 120, 121, 132, 139, 141, 146, 188, 200, 201
Cairo library, 12, 135, 201
Caliph, 5
Calligraphy, 26, 32, 36, 38, 41, 48, 54, 55
Camera, 129, 134
Cappella Palatina, Sicily, 111
Carpentry, 22
Carpet, 30, 36
Carved plaster, 55
Casablanca, 147
Castéra, Jean-Marc, 102, 217
Ceiling, 73, 75, 109, 129
Central Asia, 5, 9, 44, 56, 80, 147
Ceramic, 30, 31, 36, 41, 53, 58, 73, 75
Chelate, 103
Chester Beatty Library, 65, 106
China, 5, 9, 38
Christian, 4, 13, 18, 27
Circle-inscribed, 131, 134, 177
Collinear, 103, 145
Constantinople, 11
Conway, J., 64, 191
Copenhagen, 193
Cordova, 7, 13, 19, 54
Cosmic Equilibrium, 28
Craftsmen, 10, 34, 36, 37, 89, 93, 97, 102
Creswell, K., 217
Cromwell, Peter, 151, 217
Ctesiphon, 6

D
Dado, 146
d'Avennes, Prisse, v
Damascus, 4, 5, 10, 19, 30, 55, 140, 141
Damghan, Iran, 81
Dar al-Islam, 8
Dar-al-Hikhma, 6
Darb-e Imam shrine, Isfahan, 118, 122
Dark Ages, 22
Dart, 211
David Collection, 82
Dawla, 6
Decagonal, 117
Degeorge, Gérard, 217
Denny, Walter, 217
Derby, England, 113
Diamond, 119, 124, 139
Divrigi, Turkey, 67
Dodecagon, 147
Door, 51, 75
Dorut Tilyovat, 120

E
Ebony, 46
Edge, 105
Edge-to-edge, 76, 103, 140, 211
Edirne, Turkey, 119, 120
Egypt, 4, 7, 10, 11, 15, 17, 36, 37, 41, 43, 51, 80, 117, 147
Escher, Maurits, 69
Eski Malatya, 203
Euclid, 23, 63
Exodus, 25

F
Falsafa, 22
Fatehpur Sikri, 52, 153, 193
Fathbad, Uzbekistan, 106
Fatimid, 7, 37, 41, 51
Ferdinand, 55
Fernández-Puertas, Antonio, 217
Fez, Morocco, 54, 79, 80, 82, 84, 91, 94, 96, 99, 101, 105, 111, 113, 115, 126, 127
Formula, 71, 147, 170, 177, 181
4-way, 77, 103
France, 15
Friday mosque, Isfahan, 118, 125
Fritware, 211

G
Generalife, Granada, 128
Genghis Khan, 9, 12
Geometry, 19, 22, 32, 36, 45, 55, 63
Ghaznavid, 8, 35
Gibraltar, 5, 13
Girih, 46
Glassware, 48
Glide-reflection, 64
Gloucester, England, 111

Golden Age, 17, 22
Golombek, Lisa, 217
Gowhar Shad, Iran, 110
Grünbaum, B., 115, 218
Graeco-Islamic, 22
Granada, 14, 54
Greek, 5, 6, 17, 28
Gwalior, 202

H
Hadith, 6, 22, 27
Hafsid, 14
Harun al-Rashid, 6, 17
Hatuniye complex, Turkey, 120
Havatan Turbesi, 120
Hegira, 211
Heptagon, 131, 133
Herat, 11
Herzfeld, E., 218
Hessemer, F., 218
Hexagon, 133, 139
Hijra, 5
Hindu, 42, 51
History, 3
Holy Trinity, 28
House of Wisdom, 18, 21, 170
Huand Hatun, Turkey, 110
Hulagu, 9, 10
Humayun, 12, 52
Hungary, 11

I
Ibn Khaldun, 22
Ibn Rushd, 20
Ibn Sina, 20
Ibn Tulun, Cairo, 13, 33, 68, 188, 212
Iconography, 26
Ifrane, Morocco, 81
Ikhwan as-Safa, 20
Ilkhanid, 10
Imam al-Shafi'i, 139
Imamzada Yahya, 139
India, 19, 42, 51, 81, 117, 141, 147
Indus Valley, 4
Infinite, 27, 34
Interlace, 75, 105
Interlace, continuity, 98, 105
Interlace, discontinuity, 88, 129, 133
Interlace, strict, 105
Interstitial, 65, 68, 173, 174, 190, 213
Iran, 10, 35, 41, 42, 44, 53, 117, 139
Iraq, 9, 13, 42, 45, 51
Isabella, 55
Isfahan, Iran, 12, 53, 118, 121, 122, 124, 154, 204, 205, 207
Islamic-Indian, 12
Istanbul, 128, 199
Italy, 137
Itmad ud Daula, Agra, 52, 118, 124, 195, 197
IUC, 63, 191
Ivory, 46
Ivory-carving, 36

J
Jagy, W., 143, 169, 184, 193
Jam, Afghanistan, 81
Jamal al-Din al-Dhahabi, Cairo, 104
Jew, 4
Jihad, 3, 7
Jones, Owen, 69, 129, 218
Judeo-Christian, 25

K
Ka'ba, 25
Karatay, 59, 203
Kaseh-garan, 122
Kepler, 123
Khalifah, 3
Khargird, Iran, 112
Khatem, 39, 55, 75–78, 90, 113, 151, 183, 192, 214
Khorasan, 7, 9, 11, 35, 38
Khorezm, 10
Khwarezm, 9
Kite, 74, 87, 103, 117, 120, 141, 214
Kok Gumbaz, 120
Konya, Turkey, 59, 67, 108, 109
Kritiya House, Cairo, 146
Kühnel. Ernst, 31, 218
Kurdish, 8

L
Law, 6, 26
Lee, Tony, 218
Locher, J., 69, 218
Lu, P., 151
Lustre-ware, 44

M
Maalem Moulay Hafid, 99
Machuca, 69

Madrasah, 54, 79, 80, 94, 108, 112, 122–126, 203, 205, 207, 214
Maghreb, 44, 214
Mahdiya, 7
Mamluk, 3, 10–12, 41, 51, 135, 140, 143, 167, 201
Ma'mun, 6, 21
Marrakesh, Morocco, 54, 68, 81, 94, 95, 107, 112
Mashad al-Hussein, 56
Mashhad, Iran, 120, 121, 125
Masterpiece, 125
Materials, 41
Mathematics, 6, 17
Mawali, 6
Mecca, 5, 25
Medicine, 22
Meknes, Morocco, 54, 77, 79, 114
Menelaus, 23
Mesopotamia, 4, 43
Metalwork, 36, 41, 48
Metropolitan Museum of Art, 65, 106, 119, 120
Mihrab, 1, 29, 104, 214
Minare, 59
Minbar, 32, 46, 51, 65, 66, 105, 125, 143, 186, 199, 214
Mirador de Lindaraja, 73, 76
Mirror, 64, 139, 192
Mismatch, 92
Mitetus, Turkey, 88
Mizan, 28
Mongol, 8, 10, 22, 53, 57
Mongol-Turkic, 11
Morocco, 13, 14, 44, 53, 97, 117, 128, 140
Mosaic, 43, 53, 55, 73
Mosque, 27
Mother-of-pearl, 46
Moulay Idriss Zaouia, 91
Moustafa, A., 36, 218
Mu'awiya, 5
Mu'tazilite, 21
Mudéjar, 14, 214
Mughal, 12, 48, 49, 51
Muhammad, 3, 21, 25
Muktafi, 7
Multicoloured, 107
Muqaddimah, 23
Muslim, 4
Mutawakkil, 21
Muzzafarid, 53

N
Napoleon, 15
Nasrid, 14, 48, 54
Neal, D., 218
Necipoğlu, Gülru, 36, 155, 218
Nejjarine Palace, 99
Neoplatonic, 19, 37
Netherlands, 15
9-sided, 67
Ninevah, 6
Node, 103
North Africa, 4, 7, 139

O
Octagon, 87, 142
Ogedai, 10
O'Kane, Bernard, 120, 140, 203, 218
O'Leary, D., 17
Ornament, 105, 117
Ostia, 137
Ottoman, 11, 15, 41, 49, 56, 58, 120
Oxus, 56

P
Paccard, Andrè, 75, 97, 99, 218
Pajares-Ayuela, Paloma, 218
Pakistan, 137
Palermo, Royal Palace, 113
Paper, 37, 188
Paradise, 29
Patio de los Arrayanes, 78
Peking, 9
Pentagon, 117, 119, 128, 141, 147, 155–159, 175, 177, 199, 201
Persia, 4, 5, 9, 11, 19, 34, 41, 48, 53, 56, 137
Petal, 64, 69, 87, 109, 141, 214
Pharaonic, 29
Philosophy, 20
Photograph, 69, 105
Planar, 191
Plasterwork, 34, 36
Plato, 19
Plotinus, 20
Poet, 14
Polychrome, 44
Polygons-in-contacts, 172
Pompeii, 137
Pope, Arthur, 218
Porcelain, 68
Pottery, 48
Pythagoras, 19

Q

Qalawun, Cairo, 79
Qur'ān, 3, 6, 19, 21, 22, 26, 32, 46, 48, 120
Qutab, 51

R

Rasa'il, 21, 22
Regions, 51, 188
Registan, 57
Religion, 3, 21, 25
Revelation, 3
Rigby, John, 218
Roman, 4, 13, 17, 35, 39, 41, 69, 111, 113, 137
Rome, 6, 137
Rosette, 64, 68, 69, 73, 87, 88, 119, 120, 129, 169
Rotation, 64, 194
Roundel, 118, 193, 206, 214
Russia, 15
Rustem Pasha Mosque, Istanbul, 122

S

Sabrinye, 56
Safavid, 11, 44, 49, 53
Sakkal, Mamoun, 218
Sala de la Barca, 129
Sala de las Dos Hermanas, 73, 128
Sala de los Ajimeces, 128
Sala del Mexuar, 69, 128
Salisbury, England, 111
Salon de Comares, 62
Samanid, 57
Samarkand, 9, 11, 19, 56, 117, 122
Samarra, 33, 34
Sassanid, 4
Saunders, J., 3
Schattschneider, D., 69, 218
Science, 6, 14, 17, 21
Seherr-Thoss, Sonia, 218
Seljuk, 8, 10, 11, 35, 41, 53, 58
Serefeli, 120
Seville, Spain, 14, 88
Sevres, 68
Sfax, 137
Sgraffito, 44, 214
Shah Abbas, 12
Shahrisabz, Uzbekistan, 111, 120
Shakh-i-Zindeh, 57
Shaybanid, 57
Shia, 7, 12, 35
Sine formula, 71, 90
Singer, L., 46, 218
6-fold, 146
6-sided, 64, 67
Six-way, 52, 102, 139
Smith, Edmund, 134, 218
Spain, 4, 13, 44, 54, 117, 128
Star, 64, 117
  10-pointed, 64, 112, 119, 120, 153, 156, 160, 196, 198, 204, 205, 207
  12-pointed, 66, 67, 103, 105, 107, 109, 112, 129, 140, 142, 145–147, 169, 192
  16-pointed, 66, 68, 69, 73, 87–89, 100
  18-pointed, 67, 169
  20-pointed, 66, 128
  24-pointed, 93, 100, 147, 204, 205, 207
  3-pointed, 67
  32-pointed, 101
  4-pointed, 66, 75, 85, 95
  5-pointed, 121, 123, 131, 156, 204, 205, 207
  6-pointed, 67, 103, 105, 109, 137, 140, 142, 196, 197, 204, 205, 207
  8-pointed, 54, 66, 68–70, 73, 75, 85, 87, 88, 101, 103, 105, 109, 112, 113, 129, 192, 204, 205, 207
  9-pointed, 67, 142, 145–147
St Maria in Cosmedin, 137
Stoic, 20
Stone-carving, 34, 41
Stucco, 33, 41, 45, 48, 53, 54, 128, 188
Style, 31, 80
Sufism, 20
Suleymaniye Mosque, Istanbul, 122
Sultan Barsbay, Cairo, 90
Sultan Qaytbay, Cairo, 141
Sunni, 7, 12, 22, 35
Sura, 29
Sutton, Daud, 155, 219
Symmetry, 22, 28, 32, 45, 48, 64, 124, 191, 207
Symmetry groups, 191
  circular groups, 202
    *3• ($c3$), 204, 208
    4• ($c4$), 204, 208
    5• ($c5$), 117, 155, 205, 208
    6• ($c6$), 206, 208
    *2• ($d2$), 155, 157, 204, 208
    *4• ($d4$), 204, 208
    *5• ($d5$), 117, 157, 205, 208
    *6• ($d6$), 205, 208
    *7• ($d7$), 206, 208
    *8• ($d8$), 206, 208

*10• (*d10*), 117, 153, 207, 208
*12• (*d12*), 207, 208
frieze groups, 202
∞∞ (*p111*), 208
∞× (*p1a1*), 208
∞* (*p1m1*), 208
*∞∞ (*pm11*), 208
22∞ (*p112*), 208
*22∞ (*pma2*), 203, 208
2*∞ (*pmm2*), 203, 208
planar groups, 191
2222 (*p2*), 197, 208
*2222 (*pmm*), 76, 194, 208
2*22 (*cmm*), 199, 208
22* (*pmg*), 200, 208
22× (*pgg*), 200, 208
333 (*p3*), 196, 208
3*3 (*p31m*), 199, 208
*333 (*p3m1*), 192, 208
442 (*p4*), 194, 208
4*2 (*p4g*), 198, 208
*442 (*p4m*), 69, 147, 192, 208
632 (*p6*), 137, 139, 194, 208
*632 (*p6m*), 67, 147, 192, 208
** (*pm*), 197, 208
*× (*cm*), 201, 208
×× (*pg*), 201, 208
O (*p1*), 202, 208
Syria, 4, 7, 10, 11, 17, 32, 36, 41, 45, 51, 55, 80, 81

T
Tabbaa, Y., 36
Taifa, 13
Tamerlane, 56
Taynabiye,Damascus, 56
Tehran, 194
Tessellation, 63, 137
Tetouan, Morocco, 88
Textile, 31, 36, 41, 48
Thatta, 67
Theology, 7, 21
Thurburbo Majus, 137
Tile, 32, 83, 109, 117, 124
Timur, 10, 12
Timurid, 10, 44, 53, 57
Tipasa, 39
Tlemcen, Algeria, 192
Toledo, Spain, 13, 112
Tomb-marker, 104
Topkapı scroll, 155, 156, 197
Topkapı Museum, 125
Topology, 103
Tosh Hovli, Khiva, 139
Translation, 18, 22, 64, 211
Triangle, 71, 87, 90, 94, 119, 139–142, 147, 170, 177, 179, 192
Tulunid, 51
Tunisia, 14, 137
Turcoman, 11
Turkestan, 4, 140
Turkey, 7, 14, 21, 48, 58, 81, 117, 137, 141
Turkish Irruption, 9, 56
12-fold, 146
Two-level, 151

U
Umayyad, 5, 6, 13, 17, 28, 34, 54, 55
UNESCO, 12, 201
Unseen, 29
Uzghend, 140

V
V&A, 47, 66, 69, 121, 129
Vertex, 69, 74, 87, 97, 103, 117, 120, 124, 126, 141, 143
Vienna, 15

W
Wade, David, 215
Wakala, 140, 214
Wakala al-Ghuri, 120
Wichmann, Brian, 151
Wild, James, 121, 137
Woodwork, 30, 36, 41, 46, 51, 54, 58, 73, 75, 105

Y
Yazd, Iran, 53, 106, 198
Yazid II, 28
Yemen, 204

Z
Zangid, 36, 41
Zayanid, 14
Zellij, 94
Zoroastrianism, 29

MIX
Papier aus verantwortungsvollen Quellen
Paper from responsible sources
FSC® C105338
FSC
www.fsc.org

If you have any concerns about our products,
you can contact us on
ProductSafety@springernature.com

In case Publisher is established outside the EU,
the EU authorized representative is:
Springer Nature Customer Service Center GmbH
Europaplatz 3, 69115 Heidelberg, Germany

Printed by Libri Plureos GmbH
in Hamburg, Germany